205 - Open Pov - failed
on assumption that 6k bars provision
was key

220 - how did UF-770 i Fmc?

291 - EDL

217-8 - FOD

228 #9 ?

INTERNET AND COMPUTER BASED FAXING

The Complete Guide to Understanding and Building IP and G3 Fax Applications

Second Edition

Maury Kauffman

The Kauffman Group Inc.
Cherry Hill, NJ

www.KauffmanGroup.com

Under the auspices of Full Disclosure, the author is an enhanced fax consultant and industry analyst. Prior to this book's publication, he may have worked for or with any number of the companies mentioned herein. Also, subsequent to this book's publication, he may work for or with any number of the companies mentioned herein. Furthermore, he may invest in, or invest in organizations, which invest in, any of the companies mentioned herein.

Every effort has been made to ensure complete and accurate information in this book. However, neither the author nor the publisher can be held liable to errors in printing or inaccurate information in this book. This book is sold "as is" without warranty of any kind, expressed or implied. Readers are encouraged to verify all relevant information with suppliers when making any purchase or design designs. The author and publisher welcome suggestions and corrections.

Layout and Graphic Design by:
FrontEnd Graphics
496 North Kings Highway, Suite 231, Cherry Hill, NJ 08034

ISBN 1-57820-024-5

Dedication

With All My Love

To Nicci
Equal Partner In
The Kauffman Group

Table of Contents

Preface

In 1990, I left the frenzied world of pharmaceutical marketing to publish a newsletter, via fax. My degree in computer management told me fax broadcasting was possible, conceptually. But was it doable? I met Joe Kowalczyk, then president of Fax Interactive, one of the country's first enhanced fax service bureaus. A few weeks later, I launched *Opportunities*, a marketing newsletter for the hotel industry.

After three months, I realized selling advertising space wasn't for me. But, fax broadcasting was! Thus began the search for knowledge of all things enhanced fax.

And here I am, today. Quite luckily, writing my second book on fax technology.

"Writing" is <u>not</u> the correct verb. I am much more an editor than author. In fact, I had more help with this project than Webster had with his first edition.

The names that follow are truly champions. These are the leaders of our Industry who *really* understand what Group 1, Class 2, T.30 and X.400 mean. Not only do they appreciate this archaic, propeller-head stuff, they can form credible arguments for or against their usage. I've learned much from all of them. Without their help, **Internet and Computer Based Faxing** could not have been produced.

From the GammaLink division of Dialogic, a standing ovation to John Taylor, General Manager, from whose brain the concept and most original thinking for this book and its predecessor sprang forth. John knows more about fax technology than most everyone else in this book put together. And he is appreciated more than he knows!

Cheers to Steve Shaw, Fax Product Marketing Manager at GammaLink. Steve not only supplied the GammaLink chapters, he acts as my personal MIS consultant. He and I met several years ago in Hong Kong (of all places,) when he was in technical support. Since then he's been promoted a half a dozen times and is certainly a rising star at Dialogic. (And, I'm still just a fax consultant.)

Of course many thanks to Howard Bubb, President & CEO of Dialogic Corp, who was there, whenever I needed him. Howard's open door policy is one the rest of the industry should emulate.

In 1988, Bob Edgar founded Parity Software Development Company. Much of the telephony and voice chapters came from him and his book, PC Telephony, (which is in its third edition and can be purchased at 1-800-LIBRARY.) I highly recommend it. I thank Bob and Laura McCabe, Director of Sales & Marketing for their advice and cooperation. (Laura, if you ever need a job... you're hired!!)

The chapters explaining Brooktrout Technology's products and approaches were submitted by Andy O'Brien, VP Marketing and Business Development. When Andy started, he *was* the marketing department. Today he's successfully grown marketing to what must be dozens of people. Andy has the unique ability to appreciate all sides of an issue. Whether it's how an engineer's vision of a product could be sold or how his company fits in the many standards and strategic relations Brooktrout has joined; he understands it. This is a rare quality.

Ney Grant, founder of Ibex Technologies and today President of the Ibex division of Castelle, wrote the chapters on Ibex and Castelle products and offered tremendous advice and suggestions on updating many of the other more technical chapters. We have little in common. (He climbs mountains in Nepal and kayaks in raging floods. I hike to the nearest Marriott after I flood my basement.) Regardless, Ney trusted me before anyone else, and for that, I will always be truly grateful. Thanks!

No two guys better epitomize entrepreneurial spirit and talent than Brad Feder and Joe Cracchiolo, founders of RightFax. They did everything Right and have been justly rewarded. I congratulate them, sincerely. I also thank them both for the time and education they have given me. And of course, for their chapter. Muchos Gracias to Christie Davis for writing such an authoritative and instructional piece. (Christie, if you ever want any freelance work - call me.)

Though still a relatively small industry, the world of enhanced fax has many facets, and its experts, specialized in each, were there to help.

At Natural MicroSystems, Mike Ehrlich, Senior Product Manager and resident fax guru authored the chapter about NaturalFax. Mike, thanks for your help, speed and do-whatever-you-can attitude! I'd also like to thank Bob Schechter, Chairman & CEO of NMS for his help and ongoing industry advice.

For years, Marty Lippman has waged a (practically) one-man battle to put Commetrex on the map. And he's succeeded! Marty is everywhere there is fax, including his chapter in this book. Marty thanks..., and I'll see you at the next fax conference, and the next, and the next and the next.

When the history of fax software is written, one of the first names on the list will be Joseph Avellino, Chairman, President and CEO of Optus Software. His company has been successfully selling server software since the first fax board was manufactured. I appreciate his contribution to this effort, thank him and wish him nothing but further success.

When I decided to include a chapter on boutique enhanced fax services bureaus, there was only one person to call: Scott Edwards, president of Epigraphx. No one appreciates the sheer sales and marketing potential of enhanced fax services like Scott. The proof is how, by pushing the fax envelope, he grew his once small bureau into the nation's most sophisticated provider of fax services. Scott: thanks for your contribution and beware... you're the first person I'm calling, if I ever need a job.

At Open Port Technologies, I must thank Karen Lein for her continued support and explanations of processes I'll never understand. Also thank you to Terra Durand for authoring the Open Port chapter.

Cardiff Software and Dennis Clerke have been doing OCR longer than anyone, so if you want your faxes scanned, read and tabulated - look no further. Dennis, congratulations on your success and thanks for lending me Deborah Casey for the writing. If Casey ever needs work, all she has to do is call me.

FaxBack is the fax industries most vertically integrated company, thanks to the leadership of Art King. I'm glad he gave Kristin Roche the time to contribute chapters on Fax-On-Demand and CAS 2.0

Walter Huber was responsible for Interstar's contribution. It deserves to be here and I'm glad it is. Thanks Walt and I'll certainly see you at the next fax conference!!

A special thanks to Steve Hersey, founder and President of Copia International, for his immediate turn-around in providing one the most informative pieces on fax broadcasting I've ever read. More fax service bureaus should take his advice and his ideas!

Mark Flaherty, Product Manager authored the FaxSav chapter and assisted with several other Fax Over IP sections. Mark understands and explains the benefits of Fax Over IP much better than most. He knows what he's talking about. Thank you.

Mike Moldovan, Director of R&D at Genoa Technology got his chapter dumped in his lap. Regardless, he wrote a thoroughly educational and timely piece. An engineer who can write in English is rare. Thank you, and keep it up!

Dale Paulin has been Director of Marketing at CommercePath, since American International Fax Products sold only pulp and filament. He not only understands production fax, he's practically invented the category. I thank him for his help and submission of that chapter.

Though Omtool has been selling fax servers for nearly a decade, many people think they were lucky and came out of nowhere to command a leadership position. Insiders know that dedication and hard work, not luck was a key to their success. And Craig Randall was another key. Craig, thanks for your chapter and congrats on your success.

At Panasonic, thanks to Rich Heckelmann for his information and advice about his Internet Fax Machine. Panasonic has created and Rich is shipping what every other manufacturer has only promised, a true IP addressable fax machine.

I must also thank Glenn Josephs, CPA and partner of the accounting firm of Bagell, Josephs, Raevsky & Company, LLC. (Voorhees, NJ 609-346-2828) At a firm with a reputation as good as BJR's, that thoughtfully assists so many large and prestigious organizations, it's comforting to find a partner like Glenn who takes the time to explain and offer solutions to the complicated world of small business accounting and tax management. It's no wonder Glenn's clients include so many of the area's professional corporations. Certainly this is a main reason why BJR is one of the fastest growing and most notable accounting firms in the Delaware Valley. (Of course Carl and Allen are good too.)

And a most sincere THANK YOU to Joe Kowalczyk, CFO at NetCentric Corp. Joe, who has forgotten more about this industry than I'll ever hope to know, has been more than a mentor to me, I'm proud to call him my friend. Joe, I couldn't have gotten here without you! Thank you!

Thanks also to my family, friends and clients whom I know I've neglected, while working on this project. It is time to catch up.

Finally, thank you to Christine Kern at Telecom Books for answering my 1001 questions patiently and with understanding. I'm sorry for being such a pain.

As a consultant, I have little more than my reputation. These are good people with excellent products and services. If you need help, look no further.

If you liked this book, or didn't...
If I've made a mistake...
If I've left something out...
If you've got an idea, I need to know about....

Tell Me:
Maury Kauffman The Kauffman Group
324 Windsor Drive, Cherry Hill, NJ 08002-2426
609-482-8288 • 609-482-8940 Fax
Maury@KauffmanGroup.com
www.KauffmanGroup.com

All suggestions become the property of the author. I can't pay you, but I won't forget you, and I might make you famous!!

Happy faxing!!

Maury Kauffman, Winter, 1998

Why I Love This FAX Stuff!

Introduction

What's:

- faster and cheaper than the Postal Service?
- less intrusive than phone call?
- easier to program than a VCR?
- connects you to millions of people without bits, bauds or parity?

Facsimile!

Nothing in the world does its job faster, easier or cheaper than fax.

In an age where photocopiers are larger than automobiles and email requires 30 page manuals, the joy of fax has become one of life's simple communication pleasures.

What did you do before you were fax empowered? Can you remember, or is it blocked from memory? The effortless, elegance of fax has revolutionized communications worldwide. And, if you think the US has plenty of fax, parts of Asia (Japan and Hong Kong for example) have higher fax penetration rates than here.)

No one's sure exactly how it works... but it works, everywhere!

Through the close of the 20th Century, fax technology has become as much a part of our lives as have the telephone and electricity. We no longer pay attention to the marvel of sending our voice speeding through space, or wonder about the endless supply of power that pours from the sockets in our walls. We only notice these miracles when, for whatever reason, they do not work.

What was once considered merely a convenience for attorneys, (to discuss complicated documents,) has become a vital part of doing business. A company's very existence may hinge on whether a certain fax got through in a timely, error-free manner. Even mom-and-pop car washes depend on fax. If this seems extreme, ask yourself when was the last time you saw a business card without a fax number? (If you can remember, you're probably also noticing how young policemen, doctors, and presidents are getting.)

Since fax has become an everyday miracle, it is taken for granted. Fax is simple to install, easy to use, and reliable. You know that when the fax machine regurgitates your document and beeps, whomever you sent information to, has it. No need for a second thought.

But the times, they are a-changing...

Just as the telephone and cable services continue to evolve and offer more services and options to meet user's changing needs, so too does fax technology.

As little as ten years ago, the basic concept of faxing newsletters directly to readers wasn't even considered. Today, the US is home to dozens of enhanced fax service bureaus that send thousands of faxed pages daily. Manufacturers, VARs and System Integrators are offering computer-based fax (CBF) systems that store and forward documents and information on demand, twenty-four hours a day, 365 days per year. Individuals can now get the information they want, when they want it, automatically, using nothing more than a phone and fax.

The familiar fax machine itself is shifting and evolving. The stand-alone box is rapidly moving into the computer or the LAN. There, it offers more options, possibilities, and intriguing ways of doing business. The pressing need to use computer-based fax technologies in the international arena is opening further exciting opportunities that can be reached by this developing fax technology.

And what about the Internet. How long before sizeable volumes of fax transmissions are sent via IP, rather than the familiar G3, whether by computer or IP addressable fax machine?

However, before to you move beyond the familiar stand-alone fax machine, you must ask yourself many important questions: When do I use IP and computer-based fax? Will it be available in every PC or on the LAN? Do I standardize on modestly priced systems or do I purchase the top of the line? How do I factor in possible growth into my system? Will my system be compatible with the worldwide installed base of fax devices? And many others.

It is my hope and the hope of my many contributors that this book will answer these and other questions and help you avoid the pitfalls you may find on your way to the future.

Section I

Introduction to Internet and Computer Based Fax

Chapter 1

What is Computer-Based Fax (CBF) Processing?

Introduction

The term Computer-Based Fax (CBF) Processing means little more than shoving a fax board into a computer. However, what CBF Processing offers, is a great deal more.

Originally, fax boards were viewed simply as a way for standalone PCs to emulate fax machines. Thus, applications were limited to the transmission of ASCII files. Although fax transmission has always been used for highly visual data, poor quality reproduction was a fact of life until the advent of CBF.

CBF boards, like those manufactured by Dialogic and Brooktrout, enable users to receive faxes, print them and store them on a hard disk, as well as to create them using text and graphics files, and then transmit them. Depending on system configuration, these boards can act as fax servers for vast computer networks, sending, receiving, and routing faxes for entire multinational corporations.

A fax board is not a fax machine.

What is a fax machine?

A fax machine is a combination of three different devices: a scanner (used only when sending a fax), a modem (used for both sending and receiving), and a printer (used only when receiving a fax).

When you send a fax using a machine, a piece of paper is put into the machine. When the number is dialed, and the remote fax machine is reached, the sending machine scans the document at a resolution of roughly 100 x 200 dots per inch or 200 x 200 dots per inch (for more detailed information on the resolution of faxes, see the chapter on The Fax Image).

The scanning process changes the image on the page into digital information. This information is then sent over the telephone line via the machine's internal modem. The fax modem in the fax machine is similar to modems that you may buy for your computer, except that it only knows how to converse with other fax machines, not to data modems in computers. When the digital information is received by the receiving fax machine it is converted back into an image and printed.

So what is a fax board?

When using a fax board, the individual components of the machine are separated. The fax modem is taken out of the machine and put into a computer. To emulate a fax machine, the computer must be relied on to scan (or otherwise convert) documents into fax format.

On received faxes, the computer must also provide the equivalent of printing faxes, which also may be displaying them. So by now you may be asking yourself, if a computer based fax board is a fax machine minus a printer and minus a scanner, why do they sometimes cost more than an entire machine? The reason is software.

In the case of the fax machine, the only software that is needed is internal software to provide the 3 basic functions listed above. A fax board may contain a lot more computer related software to provide the glue that allows a user to convert documents into fax format, to print and display faxes, to manage the queue of outgoing faxes, to keep track of what faxes went to which numbers, to keep track of which faxes failed and for what reason, to manage more than one board in a computer at the same time, etc. In essence, once the process is computerized, a lot more information related to the faxing tasks can be tracked and controlled.

There are many other advantages to CBF Processing. They include:

- Higher Quality Output
- Labor Saving
- Convenience
- Call Progress Monitoring
- Confidentiality
- Efficient use of Telephone Lines
- Reduced Telephone Expenses

Higher quality output

Computer based fax can produce much higher quality fax pages than the normal quality that is received when faxes are sent via a fax machine. The scanner in a fax machine is subject to inaccuracies due to the paper not being perfectly aligned in the machine and round-off error in the scanning process. With a fax board, the computer knows that the output will be printed on a fax machine at fax resolutions, so it can much more intelligently convert documents into an image that is very readable by the recipient of the fax. Some of the advanced boards on the market support conversion from advanced page description languages such as PostScript and PCL. When documents are converted from these page description languages directly into fax format, the results can be stunning.

Labor Savings

The labor saving is tremendous, especially if sending the same or very similar document to more than one different person. Computer based fax is great labor saving device if any sort of repetitious faxing is done. For example, many marketing departments have data sheets and price lists on their products that they frequently send out to prospective customers. Rather than having a secretary spend a portion of each day sending the same information to different fax numbers, this can easily be done via the computer. The customer specific information such as telephone number, fax number, and address is taken one time. Once the data is in the computer, the customer can receive any documents that are stored on the computer. Another variation on this theme is to have customers call up a computer and choose among documents that have been previously stored on the computer. These systems are known as fax-on-demand systems and are described in more detail in Section VII.

Convenience

For those documents created or stored in computer systems.

This benefit is similar to the labor saving benefit above. But in this case, the reference is to any document that is created on the computer. To print out a document, then take that same document and feed it into a fax machine is a cumbersome process that can be streamlined by sending the document directly from a computer. For example, let's say that I want to send a message about an upcoming meeting to five individuals, 3 of whom are in my company and 2 who are not. I get on my email system and send an email to the 3 who are in my company. Since I have a sophisticated CBF system, I just type in the names and fax numbers of the 2 individuals who are not in the company and the same message goes to them as well. This is much easier than printing out the email message and sending it by hand.

Call Progress Monitoring

Computer based fax can monitor the progress of each call and handle each fax task differently depending on the results of the first call. For example, let's say that I want to send a newsletter to 50 subscribers. On the first attempt to send the newsletter, 30 of the faxes go through the first time, 10 of the faxes encounter a busy signal, 5 get a voice on the line, 2 receive a paper out indication, on 2 there is no answer, and on the last one there is a message that the number has changed. For the busy faxes, I want to try to send again in a short amount of time, say 5 minutes. For the voice numbers, I don't want to try the fax again, but alert the sender that the numbers are not fax numbers and that they should be changed. For the 2 numbers where I receive paper out indication, I want to try again at one hour intervals, hoping that the user has replenished the paper by then. I may need to continue trying to send for several days. On the 2 that receive no answer, I will probably want to try again at a long interval (maybe the user accidentally unplugged the machine from the telephone line). If I am not successful after a predetermined time interval (say a day or two), then I should let the sender know that no one answers at these numbers. One the last number, where the number has been changed, I don't want to try sending again, but notify the sender that the number has changed, and that he should call the number to find out if there is a new number. Thus, by monitoring the progress of each call, the computer based fax system allows the sender to take much of the labor out the process of sending to machines that are not always 100% available.

Confidentiality

Receiving faxes via fax machine can compromise the confidentiality of incoming faxes in many cases because the fax machine is in a public area. Computer based fax can address this problem in several different ways. One way is to install a LANFax server and give every recipient their own DID fax number. Each fax goes directly to the user and only the user determines when to print out his faxes or view them confidentially on his screen. Another way is to use fax mailboxes. A fax mailbox is a fax number that holds all of a user's faxes until the users is ready to print them out. The user can retrieve faxes while traveling or can go to any fax machine and retrieve them without worry about someone else seeing the faxes that were intended only for him.

Efficient use of Telephone Lines

Or, better stated, more efficient use of expensive telephone lines.

In larger organizations, many users may have the need to send or receive faxes. Once they see the benefits of computer based fax, they all may want to participate and ask for a second telephone line to be hooked up to their computer for the occasional fax that they need to send. Having a separate line for each user can be much more expensive that by concentrating the fax traffic onto a much smaller number lines (a fax server). By funneling all fax traffic through a much lower number of telephone lines, the monthly charge for basic telephone service can be reduced dramatically.

Reduced Telephone Expenses

Some of the leading fax boards on the market support the most advanced compressions and line speeds that are present in the installed base of fax machines. (For more information on fax compression, see chapter 7). This can make a large difference in a long distance telephone bill if there are a lot of faxes being sent long distances, especially internationally. 30 to 50% savings are possible if fastest speeds and densest compressions are used.

Chapter 2

Why Internet Fax

or Fax Over IP (FoIP)?

Introduction

During the past decade, growth of the Internet has been explosive. In 1986, 5000 computers were connected via the Internet. By 1991, the number of computers on the Internet had climbed to 500,000. Today, nearly 10 million computers are connected, and the expansion shows no signs of slowing down. Forecasts predict that 100 million computers will be connected by the end of 1999.

Fax messages are ideal candidates for transport over the Internet. Faxes are simply digital images. Because the Internet is designed to send digital data, no degradation of the message occurs with sending, routing, or forwarding.

Corporate Fax Trends

Fax is a vital means of business communication. Ease of use, international compatibility, and immediate delivery have contributed to the worldwide acceptance of fax. Sixty million fax machines are currently in service, and the market continues to grow. Globally, annual sales of fax machines reached more than 12 million units in 1997, and estimates call for 20 million annual units by 1999.

While computers have contributed to dramatic changes in workplace processes, fax continues to have strong appeal in offices around the world. For example, although the vast majority of faxed documents originate from computers, nearly 90% of Fortune 500 fax users send faxes using standalone fax machines. Surveys shed some light on users' rationale. According to the Gallup/Pitney Bowes Fax Usage and Application Study, fax is perceived as more reliable and more likely to generate a faster response than email, voice mail, or overnight courier. Users also find that printed fax documents are easier to read. In spite of available electronic alternatives, most still prefer hard copies for incoming documents. The preference for paper is understandable. Few computer monitors can effectively display a full-size page of text. It is not a coincidence that computers have emulated the size, shape, and weight of a notebook. Paper continues to be an efficient form of communication.

Sending Fax over the Internet

Today, fax messages are sent via the Public Switched Telephone Network, or PSTN. Calls are billed per minute, incurring the highest charges during the business day. These same fax messages can be delivered using the Internet at a significant savings. By using the Internet to deliver fax traffic, documents can be delivered without the need for long distance calls. The idea of merging fax traffic and computer networks is not new. What is revolutionary, however, is the capability to deliver fax messages internationally without incurring long distance telephone charges.

To illustrate, consider the following scenario. A corporation needs to fax a document overseas using its local area network (LAN) fax server. Normally, the document is sent to the LANFax server, which initiates an international phone call to the destination fax machine. This same exchange could happen across the Internet, but without the need for an expensive international call. Just as before, the user sends the document to the fax server. In this case, the fax server does not dial the phone number; rather, it uses the Internet to contact an Internet fax gateway near the destination fax machine. The fax is transmitted across the Internet to the remote fax gateway. The remote fax gateway initiates a local phone call to deliver the fax. In this manner, fax documents can be delivered to any destination, international or domestic, with only local calls.

Internet delivery is not limited to faxes originating on computers. Standalone fax machines can transmit across the Internet, too. Currently, fax machines are connected directly to the telephone network. Phone numbers let various local and long distance carriers initiate connections and deliver faxes. To accomplish this delivery across the Internet, fax servers could collect traffic from standalone fax machines and route it to the Internet for delivery. From the user's perspective, nothing has changed. Just as before, the user dials the destination phone number, but the Internet fax server interprets the dialed digits, not the PSTN. The fax server would use the dialed digits to locate an appropriate remote server and initiate an Internet connection. Next, the fax would be transmitted via the Internet to the remote fax server. Document delivery requires only a local phone call when the destination is near an Internet point of presence. Using methods such as these, the Internet becomes a viable alternative to the PSTN for the delivery of fax documents.

Why Internet Fax is Attractive

Communication costs represent a sizable corporate expense. Worldwide, fax machines accrue $30 billion in annual phone bills. At the typical Fortune 500 company, 37% of the phone bill is attributable to fax traffic. When all costs of fax ownership are considered, communication charges are the most significant, exceeding the purchase cost of the machine itself.

Phone companies charge for every minute of use. Internet use is typically unmetered; that is, the price of use is the same regardless of the destination or volume of traffic. By routing fax messages across the Internet, companies can avoid the expense of long distance and international toll calls. The potential for savings is enormous.

While not all fax machines are near an Internet point of presence, Internet fax routing still can offer significant cost savings. For faxes which must be completed using toll calls, least-cost routing algorithms are being developed to dynamically route fax traffic through the most economical path. Often traffic will be routed to a server in close geographical proximity to the destination, but other options are conceivable. For example, a fax from New York to Beijing may be sent over the Internet to Sydney to take advantage of midnight phone rates. With points of presence in multiple time zones, traffic can be sent via the Internet to a location where phone tariffs are least expensive. In this manner, routing fax over the Internet has the potential to reduce communication costs even when the destination is not connected to the Internet.

Intranets - Company Internets

Many companies have already invested in high-capacity internal networks, or intranets, to carry corporate data traffic. Intranets operate identically to the Internet, but are dedicated to company functions. Often, intranets are linked to the global Internet, but it is not necessary for internal traffic to pass onto the Internet. Instead, corporations lease dedicated connections between distributed offices. In fact, 72% of Fortune 500 companies have implemented such networks. Once an infrastructure with sufficient bandwidth is in place, the cost to send additional traffic through the network is virtually zero.

According to the Gallup/Pitney Bowes Fax Usage and Application Study, nearly half (48%) of Fortune 500 fax transmissions are from one company location to another. By sending fax traffic via corporate networks rather than the telephone network, a company can reduce internal fax transmission costs to virtually zero. Further, least-cost routing ensures that external fax communication costs are kept to a minimum as well. Least-cost routing lets fax messages travel across corporate networks before being transmitted over the PSTN. A fax can be initiated in New York and sent to the London office over the company network. No long distance charges are incurred. In London, the company's intranet fax server can deliver the fax to the final destination. Previously, an international call would have been necessary.

Internet Fax Applications

Transmission of fax traffic over the Internet requires the integration of fax with the computing and networking infrastructure. Integrating computers, networks, and fax machines creates numerous opportunities to enhance fax delivery and reduce fax expenses. Several implementations of Internet fax are described below.

Email to Fax Gateways

The earliest union of fax and the Internet let email messages be directed to fax machines. Using these applications, email users can address email recipients and fax recipients in one message. Simple text messages are supported, as well as TIFF files and MIME-compatible attachments. With the rapid growth of Internet users, this type of service has become even more attractive.

Internet Fax Gateways

Building on the email-to-fax gateway, advanced gateways will route traffic both onto and off of the Internet, and serve as the building block for a fax-enabled Internet. Gateways will be deployed in two configurations. As customer premises equipment (CPE), on-site gateways will let a company send fax traffic over internal networks and the Internet. Second, Internet fax gateways will be deployed at communications centers, such as Internet Fax Providers (IFPs), fax service bureaus, or telecommunications companies.

At a communications center, fax delivery will be provided as a service, using the Internet rather than the PSTN for delivery. Ideally, communication between gateways will be based on open standards, allowing interoperability and reliable exchange of fax messages. In each circumstance, least-cost routing will ensure that communication costs are minimized. Internet fax gateways will be accessible to stand-alone fax machines, as well as PC fax clients, letting all fax traffic be delivered via the Internet.

Universal In-box

The task of managing messages is time consuming, often involving disparate processes to retrieve email, voicemail, and fax documents. In this area, convergence promises to have a significant impact on workflow. Tools have been developed to manage fax, email and voice-mail from a single application, eliminating the need to monitor multiple systems.

The universal in-box can be a powerful tool for mobile users as well. Using the Internet for retrieval, all messages—fax, email, or voice—can be delivered anywhere in the world. A destination gateway stores messages until a user needs to retrieve them. Users connect to the gateway via the Internet. Messages are then downloaded to the user. Access to the Internet is available virtually world-wide through numerous Internet Service Providers (ISPs). In addition, many companies provide dial-in network access for employee use. Once connected to the Internet, several retrieval methods are conceivable, including multimedia email and protocols supported on the World Wide Web.

WWW/Fax-on-Demand Server

In spite of the phenomenal growth of the Internet, usage is not yet universal. One potential integration of fax and the Internet makes information on WWW pages accessible from fax machines. Fax on-demand servers with Internet capabilities can retrieve web pages based on touchtone selections. In this manner, information providers can maintain data on a web server, while distributing that information to the wide audience of users who have access to fax machines, but not full Internet access. By enabling the faxing of HTML documents, new generations of fax-on-demand servers let companies have a single store of documents that can be accessed by whichever method (WWW or fax) that users find convenient.

Extending the Capabilities of Standalone Fax Machines

High-end fax machines typically boast a rich set of features, such as broadcasting and usage reporting. Some even let the fax function as a networked scanner and printer. Many advanced capabilities can be provided to the simplest fax machine by implementing these features in an Internet fax gateway.

Following are just a few of the advanced features that can be supported with Internet fax.

Fax Usage Reporting and Management

One-quarter of companies do not track fax costs, yet 41% of telecom managers expect fax-related telephone charges to increase in the next year. Once management knows which sectors of the company spend the most for fax, when the majority of faxes are being sent, and how much is being spent, it becomes possible to control these expenses. Typically, these costs are buried in regular telephone reports, making it difficult to discriminate between fax and voice use. By routing fax traffic through servers and the corporate network, usage is easily monitored.

Secure Delivery

Information security is a serious concern regardless of the method used to send documents. While much concern is directed to the issue of Internet security, Internet fax delivery has the potential to increase document security. Fax transmissions over normal phone lines are transmitted in compressed digital form, and can easily be deciphered. Further, fax machines are often shared, and documents can be read by any number of passers-by. Internet fax routing can provide delivery directly to a recipient's private in-box. Access can be password restricted. For truly private messages, public key cryptography can transform a digital fax message into an indecipherable encoded message for delivery across the network. Only the intended recipient can decode the fax for viewing.

Guaranteed Delivery and Never-busy Receipt

When a fax is sent via the Internet, delivery options are varied. If the destination phone number is busy, the fax message could be received and stored for later delivery. The fax server could manage any redialing, and provide information to the sender regarding the delivery status.

Fax Broadcast

Frequently, the same document must be delivered to multiple recipients. This can be a tedious process on many fax machines. When connected to an Internet fax gateway, additional capabilities are available. For example, distribution lists can be maintained and stored on the server, simplifying the task of broadcasting fax messages.

Obstacles or Opportunities?

High-Quality Service

Fax has flourished due to many favorable characteristics. Fax is easy to use, and machines can communicate internationally. Virtually any paper document can be sent immediately. For fax over the Internet to thrive, the features that have brought widespread acceptance to fax should be preserved as much as possible. Modifying the service quality without providing additional features would deter the acceptance of Internet fax delivery. Further, any implementation of Internet fax must be capable of seamlessly communicating with the installed base of standalone fax machines.

Currently, fax transmission reliability is very high. At a minimum, Internet fax should match the reliability of delivering fax over the PSTN. Internet fax has the potential to deliver higher levels of service through extended services such as guaranteed delivery.

Phone/IP Translation Infrastructure

Users find fax messaging extremely easy to use. Ideally, users will continue to interact in exactly the same manner as always, that is, a simple phone number initiates a connection. Simplicity without sacrificing features will increase the likelihood of adoption.

Internet fax can be integrated transparently into office environments provided that the method to identify a destination continues to be a common phone number. To accomplish this, a method must be devised to map telephone numbers to Internet destinations. Successful experiments have proven the technical feasibility of sending faxes across the Internet and mapping international phone numbers to the IP addresses of networked fax servers.

Standards

Standards to exchange fax messages across the Internet are just emerging. Common industry standards will increase interoperability. As interoperability increases, so does the value of routing faxes over the Internet, leading to more rapid implementation and greater penetration.

Standards are critical for the growth of this market.

Why Develop Internet Fax Solutions?

End User Benefits

It is clear that Internet fax can provide significant cost savings to users. Additional services such as guaranteed delivery and enhanced security provide still more value to fax users. Considering the potential cost savings, the widespread usage of fax and the dramatic increase in Internet use, Internet fax is poised to grow rapidly.

Internet Fax: An Opportunity for ISPs

The benefits of Internet fax will create additional demand for TCP/IP network services, resulting in additional customers for ISPs. Potentially, advanced messaging features such as Internet fax could be billed per transaction or packaged as a separate service, providing new revenue opportunities for ISPs. Some ISPs have developed world-wide network coverage, and are well positioned to provide worldwide fax delivery via the Internet. As standards permeate the market, smaller ISPs can participate as an Internet fax point of presence, providing additional coverage for fax delivery.

Opportunity for Fax Server Developers and Resellers

The Internet is changing the way businesses exchange information. Businesses are continually searching for technologies and processes to increase efficiency, and the Internet is a powerful foundation on which to build. Solutions that are Internet enabled have a distinct advantage vs. products with no Internet capability. Internet delivery is not a replacement for fax servers, but an enhanced capability that complements existing products. Customers already enjoying the benefits of a LANFax server would receive additional benefits from Internet delivery, through enhanced delivery options and reduced telecommunications costs.

This chapter was adapted from a GammaLink whitepaper on Internet Fax, available at: www.gammalink.com

Chapter 3

The History of Fax

The concept of sending electrically a reproduction of an image to a distant recipient is almost as old as the telegraph itself; in fact, it could be said they were almost born together. In those days, fax used telegraph technology to transmit: a low-speed direct current (DC) line with a single wire to ground, and no amplifiers other than slow electrical relays which curtailed the distance which these signals could be sent across. Both telegraph and fax sent their information by the same means: contact switching of metal patterns and interrupted current, provided by wet cell batteries, to send off the information. Both received by marking on paper.

The first successful fax was patented in 1843 by a Scottish inventor, Alexander Bain. His Recording Telegraph worked over a telegraph line, using electromagnetically controlled pendulums for both a driving mechanism and timing. At the sending end, a stylus swept across a block of metal type, providing contact scanning wherever the type stood out from the block. This caused a voltage to be applied to a similar stylus at the receiving end, reproducing an arc of the image on a block holding a paper saturated with an electrolytic solution which discolored wherever an electric current was applied through it.

The blocks at both ends were lowered a fraction of an inch after each pendulum sweep until the image was completed. Bain's device transmitted strictly black and white images: it was unable to produce a scale of grays in between, but almost two centuries ago this was not of too vital importance, compared to the capability to send an image (regardless of how poor) over telegraph wires! Soon after Bain's invention, several versions of his idea found relatively wide application.

The first commercial fax service was started in 1865 by Giovanni Casselli, using his Pantelegraph machine, with a circuit between Paris and Lyon, which was later extended to other cities. By the 1930's, systems using photoelectric sensors and rotating drums were commonplace in newspaper offices and law enforcement agencies to send and receive photographs and other graphic material over telephone wires.

The first versions of what we would recognize as an electronic (rather than purely electromechanic) fax machine used photoelectric tubes to measure the brightness of each spot on a document's surface as it was quickly rotated on a drum. This allowed it to transmit gray-scale information. These strictly analog devices worked basically by producing a varying voltage based on the output of the photoelectric tube, with the process being reversed at the receiving end.

A problem with this system was that since there might be variations in brightness at a frequency below that of the audio range of telephone lines, the public switched telephone network could not be used. DC-coupled leased-line circuits were necessary.

Amplitude modulation (AM) was used to solve the problem of low frequencies. The varying brightness was no longer used to vary the voltage on a line but, instead, to modulate an AM carrier. Although the concept worked well, it was extremely sensitive to line noise and if circuit gain changed during transmission the image was distorted (sometimes catastrophically) by light and dark bands. Specially conditioned leased lines were still a requirement.

Other modulation schemes were tried and used: frequency modulation (FM), phase modulation (PM), and vestigial sideband (VSB) modulation, the latter a form of AM that compresses required bandwidth. After the Second World War there was great interest on the part of newspapers in using fax technology to send newspapers directly to subscribers' homes, but the coming of television as well as technical problems forced an abandonment of the idea.

We are accustomed to the machines which appeared during the course of the last decade, when scanners and thermal printers did away with spinning drums, making fax technology available to those outside of journalism, the military, and law enforcement. Businesses were the first to profit and, in increasing quantities, personal users. But before this could happen the Babel of languages these machines used, which kept them from communicating with each other, somehow had to be brought under control.

The Coming of Standards

It was not until October 1966 that the Electronic Industries Association proclaimed the first fax standard: EIA Standard RS-328, Message Facsimile Equipment for Operation on Switched Voice Facilities Using Data Communication Equipment. This Group 1 standard as it later became known, made possible the more generalized business use of fax. Although Group 1 provided compatibility between fax units outside North America, those within this region still could not communicate with other manufacturers' units or with Group 1 machines. Transmission was analog, it typically it took between four to six minutes to transmit a page, and resolution was very poor.

U.S. manufacturers continued making improvements in resolution and speed, touting the three-minute fax. However, the major manufacturers, Xerox and Graphic Sciences, still used different modulation schemes FM and AM, again, there were no standards. Then, in 1978, the now ITU-T came out with its Group 2 recommendation, which was unanimously adopted by all companies. Fax had now achieved worldwide compatibility and this, in turn, led to a more generalized use of fax machines by business and government, leading to a lowering in the price of these units.

When the Group 3 standard made its appearance in 1980, fax started well on its way to becoming the everyday tool it is now. This digital fax standard opened the door to reliable high-speed transmission over ordinary telephone lines. Coupled to the drop in the price of modems, an essential component of fax machines, the Group 3 standard made possible today's reasonably priced, familiar desk top unit.

The advantages of Group 3 are many; however, the ones that quickly come to mind are its flexibility, which has stimulated competition among manufacturers by allowing them to offer different features on their machines and still conform to the standard. The improvement in resolution has also been a factor. Similarly to the regular television set, fax clarity or resolution depends on the number of lines present in the fax: the more lines, the clearer the image is. The standard resolution of 203 lines per inch horizontally and 98 lines per inch vertically produces very acceptable copy for most purposes. The optional fine vertical resolution of 196 lines per inch improves the readability of smaller text or complex graphic material.

Another factor has been transmission speed. Group 3 fax machines are faster. After an initial 15-second handshake that is not repeated, they can send an average page of text in 30 seconds or less. Memory storage features can reduce broadcast time even more. The new machines also offer simplicity of operation, truly universal compatibility, and work over regular analog telephone lines, adapting themselves to the performance characteristics of a line by varying transmission speed downward from 14400 bps if the situation requires it.

Modulation is the process of varying some characteristics of an electrical carrier wave (CW) as the information to be transmitted on that CW varies. The three most used kinds of modulation commonly used for communications are AM, FM, and PM.

The Telecommunications Standardization Sector (TSS) is one of four permanent parts of the International Telecommunications Union (ITU), based in Geneva, Switzerland. It issues recommendations for standards applying to modems and other areas. Although it has no power of enforcement, the standards it recommends are generally accepted and adopted by industry. Until 1993, the TSS was known as the Consultative Committee for International Telephone and Telegraph (CCITT). It is now referred to as the ITU-T.

Section II

Understanding Basic Telephony

Chapter 4

The Complete Phone Call:

Off-hook to On-hook

Introduction: The Phone Line

Most domestic telephones are connected to the telephone company's nearest exchange using a cable containing two conducting wires.

The telephone company exchange is called a Central Office or simply CO. The CO is similar to a business phone system but on a much larger scale. Phone systems and other devices which can connect calls are called switches. Your phone is therefore connected directly to the CO switch. Business phone systems which function like CO switches are called Private Branch Exchanges, or PBXs (sometimes PABX, with an extra "A" for "Automatic").

The connection to the phone company using two wires is called a two-wire connection. To distinguish it from other types of connection, it may also be described as analog, since sound is represented by varying current rather than by digital signals. A regular business or domestic phone line without special features may also be called a POTS line, for Plain Old Telephone Service. A line to the phone company lets you connect to a number anywhere in the world through the Public Switched Telephone Network, or PSTN.

Some business phone systems (PBXs) use more than two wires to connect a phone on a desk (the station set) to the PBX. The extra wires are used to send signals between the station set and PBX, which can be used to implement message waiting lights, LED displays, conferencing and other features. A "standard" voice card will not support the additional wires, and uses a connection to the PBX which is like a domestic wall socket, requiring a two wire analog station card in the PBX. The CO provides a small DC voltage across the two wires, called (for obvious reasons) battery.

The two wires are known as tip (often connected to battery -) and ring (often connected to battery +). For most purposes, it doesn't matter which way round tip and ring go, but it is wise to get it right anyway.

Starting An Outgoing Call: Getting Dial Tone

When the hand-set is taken out of its cradle, the phone is said to be off-hook. The action of taking the phone off-hook closes the connection AB, so that there is a complete circuit to the CO along the local loop. This causes current to flow, known as loop current. The CO switch will usually react to this by making a sound (a combination of 350Hz and 440Hz tones), known as dial tone which indicates that you may dial.

Most analog lines are loop start, which request dial tone in this way C the alternative is ground start, where service is requested by grounding one of the two conductors in the two-wire loop.

Announcing An Incoming Call: Ringing

When the hand-set is in its cradle, the phone is said to be on-hook. When the phone is on-hook, the connection AB is broken, but there is still a complete circuit made through a capacitor, shown as 444 in the diagram. When a call arrives, the CO applies an A/C voltage of about 105V to the circuit (ring voltage), and the phone rings.

Dialing A Number

There are two fundamentally different ways of dialing numbers: tone dialing and pulse dialing. Pulse dialing is sometimes called rotary dialing because that is the method used by old-style rotary phones.

Tone dialing uses sounds to represent digits (we include 0 through 9, # and * as digits). Each digit is assigned a unique pair of frequencies, hence the name Dual Tone Multi Frequency (DTMF) digits, or Touch Tone digits. There are four DTMF digits (in the fourth column C because the usual tone pad has three) which are not usually found on telephones in the US but are used in some European countries, these are named a, b, c and d.

There are actually two standards for tone digits: DTMF and MF (for Multi Frequency). MF is used internally by the phone company, but is also used by some phone companies for ANI (Automatic Number Identification) and occasionally for other services. ANI is the business equivalent of Caller Identification, where the telephone number of the calling party is transmitted to the telephone receiving the call. MF is very similar to DTMF except that different pairs of frequencies are used for each digit. Some speech cards have the ability to detect MF as well as DTMF, others, especially older models, do not.

Pulse dialing uses the loop current itself to send digits. When the dial of a rotary telephone rotates, it briefly turns the switch AB on and off, thus turning the loop current on and off, resulting in "pulses" of loop current. Count one for each pulse to get the digit being dialed. You know when a digit is finished and the next one starts by the longer pause between pulses.

Pulse dialing has two major disadvantages. The first is that pulse is much slower than tone dialing, the second is that most switches will not transmit pulses over a connection. If you make a call from a rotary phone, and dial a pulse digit in the middle of the conversation, clicks will be heard at the far end, but no interruptions will be made in the loop current. This means that the only way to detect pulse digits from a remote telephone is to try to analyze the sound patterns and to "guess" when a digit has arrived. Imagine distinguishing a pulse "1" digit from the click caused by static on the line, for example C not an easy problem. Systems capable of recognizing pulse digits from the sound they make are called pulse to tone converters.

Completing The Call

When dialing is complete, the person or equipment which dialed the number can listen to the line to determine when and if the call is completed, i.e. if the called party answers the phone. Along the way, a number of call progress signals may be generated to indicate how things are going in the process. Call progress signals are mostly sounds (tones) generated by a switch, some signals are made by dropping loop current briefly.

Ringing tones (called ring-back in the business), indicate that ring voltage is being applied to the line corresponding to the number dialed. Ring-back is generated by the CO switch which is attached to the number that you called, not by the called phone (there may be no equipment attached to the number at all).

If the dialed number is off-hook when the connection is attempted, a busy signal will be generated instead. If the phone company's network is busy and the local CO (the CO you are attached to) fails to make a connection to the distant CO (the CO connected to the dialed number), a fast busy may be generated. You don't hear fast busy tones very often C they sound similar to busy, but the pause between the beeps is shorter.

If you dial a bad number (an area code that does not exist, or a disconnected number), you will get an operator intercept signal (three rising tones) followed by a recording: "doo-doo-doo We are sorry, ..."

Less obvious but still important are brief drops in loop current which are sometimes generated when making long-distance calls. These can be used by the phone company as an acknowledgement that the distant CO has been reached, and are often used to indicate that the called number went off-hook. It is all too easy to confuse these brief loop current drops with the drop in loop current that signals a disconnect (end of a call due to the called party hanging up). Finally, if all goes well, the called party will answer the phone and say "hello."

Call Progress Analysis

Call progress addresses the important and difficult question: what is the state of the telephone connection?

Call progress analysis, sometimes known as call supervision, is most commonly performed immediately following the dialing process which communicates the desired routing of a call to a switch. The results of the attempt are indicated to the calling person or equipment originating the call in the form of call progress tones. Common examples include ringing tones and busy signals.

Call progress analysis presents a difficult challenge to automated equipment such as voice processing hardware because of the antiquated technology used in signaling. The most commonly used telephone connection, the analog two-wire interface, provides no separate signaling channel C all call progress signaling must be done using in-band signals. In other words, the only means available of signaling the progress of a call is to use the sound (audio) on the line, or the loop current. More modern digital connections such as ISDN and the European E-1 standard provide for out-of-band signaling, where signals completely separate from the audio channel can be used.

While some switches do provide loop current signals to indicate the completion of a call, in most cases the voice processing hardware will need to analyze the received audio on the line to match the sound against expected patterns of tones. This process is complex, and requires a great deal of tuning and flexibility if it is to be completely robust, especially when many different environments may be encountered. For example, a voice mail unit may be required to transfer a call within a PBX, dial pager units through local and long-distance calls, or forward voice mail messages over the public network to other national or international offices. It is a major challenge to provide reliable call progress features for all these different environments. Each PBX will have its own characteristic tones for ring and busy, these may even vary between different models from the same manufacturer.

A typical call progress algorithm reports the results of call progress analysis as one of the following:

Ring no answer	Ringing tones were detected, but after a pre-set number of rings, there was still no answer.
Busy	A busy tone was detected.
Fast busy	A fast busy tone, indicating that the network was unable to reach the desired CO, was detected.
Operator Intercept	The "doo-doo-doo We are sorry..." tones were detected.
No ring-back	After waiting for a set length of time, ringing tones were not detected.
Connect	The call was answered.

Call Progress Monitoring

The term call progress analysis or supervision is usually applied to the period immediately following an out-dial or transfer. A related but more general feature is call progress monitoring, which refers to the entire length of the call. The simplest, and generally most reliable, aspect of call progress monitoring is hang-up, or disconnect, supervision.

In analog environments, hang-up supervision is usually implemented by dropping loop current for a brief period. Digital environments will generally signal a disconnect by changing the value of a signaling bit. On standard T-1 circuits, for example, the A bit transmitted to a device will generally be changed from A=1 (corresponding to loop current active) to A=0 (corresponding to loop current absent).

The bad news is that some switches, including unfortunately many business phone systems, do not provide loop current disconnect supervision. To detect a disconnect in such an environment, it may be necessary to monitor the line continuously for the tone generated by the switch when the call has been terminated and the line is left on hook C generally a new dial tone, or a re-order tone (beeps, possibly followed by a message like "Please hang up and try dialing your number again").

The Flash-Hook

When a call is in progress, a service from a CO, PBX or other switch may be requested by making a flash-hook. A flash-hook puts the phone on-hook briefly C long enough for the switch to detect it, but not long enough to disconnect the call. You will probably be familiar with a flash-hook from the domestic Call Waiting feature.

On a business phone system, a flash-hook will generally give you a second dial tone, allowing you to make a three-way conference (you stay on the line, flash-hook a second time to complete the conference), or a transfer (you hang up when the second number answers).

Most local phone companies offer Centrex features, which allow transfers and three-way conferencing using the same flash-hook and dial sequence as a typical small business phone system.

Terminating A Call: Disconnect

To terminate a call, one end goes on-hook, ie. hangs up the phone. Sooner or later (there may be a delay of twenty seconds or more, less for a local call), the phone company will transmit the disconnect by dropping the loop current. Some phone systems don't transmit a disconnect at all, so you may have no way of detecting the end of the call at the local end.

This chapter was adapted from Bob Edgar's book: PC Telephony, available from Flatiron Press, (800-LIBRARY.) I thank Bob, founder of Parity Software Development Corporation for his help and advice with this chapter. I recommend his book wholeheartedly.

Chapter 5

Call Routing

Introduction

There are various ways inbound telephone calls can be routed. Description of several basic methods follow.

DNIS And DID Services

Many phone companies offer a service which allows you to terminate several different phone numbers on the same set of lines. When an in-coming call arrives, it searches (hunts) for an available line and generates a ring signal. Your equipment responds with a wink signal, which causes the phone company switch to send you some or all of the digits that the caller dialed, usually as DTMF digits. This is called Dialed Number Identification Service, or DNIS (pronounced "dee-niss"). A T-1 wink is a brief off-hook period, which will usually be signaled using the A bit on a T-1 channel.

DNIS is used by answering services and service bureaus, which may be answering calls from many more numbers than they have operators or IVR lines.

An analog wink is done by reversing battery voltage. Analog DNIS is called DID, for Direct Inward Dial, and sometimes DDI. A typical use of DID is to provide direct numbers to employees in a large company. The company phone system obtains the last four digits of the dialed number via DID and routes the call to the appropriate extension.

To get DNIS or DID digits and answer a call, you would use a sequence like this:

1. Wait for ring
2. Send wink
3. Get DNIS digits
4. Go off-hook

T-1 Digital Trunks

A T-1 digital trunk carries 24 telephone connections on two twisted pair (two-wire) cables. The set of 24 connections is called a T-1 span. Each of the 24 connections is referred to as a time-slot. Each time-slot carries sound, digitized at 64Kbps, and two signaling bits, referred to as the A and B bits, which play a role similar to loop current signaling on analog lines. One time-slot is sometimes referred to as a DS-0, for Digital Signal level 0, signal.

While use of the A and B bits is not the same in all equipment, a common convention, called E&M signaling, keeps the A and B bits equal, and uses the A bit to indicate whether or not a connection is active. Thus, A bit high (set to 1) corresponds to loop current flowing, A bit low (set to 0) corresponds to no loop current.

Digits can be dialed on a time slot using the same methods as analog lines: DTMF, MF and pulse. Pulse digits are sent by turning the A and B bits on and off, just as rotary pulse dialing turns loop current on and off on an analog line. With E&M signaling, a pulse would be sent by briefly changing the off-hook state (AB = 11) to on-hook (AB = 00) and back again.

The complete T-1 span with 24 DS-0 signals and synchronization information is called a DS-1 signal.

Automatic Number Identification (ANI)

ANI (pronounced "Annie") delivers the phone number of the person (or machine) calling. It functions in a similar way to DNIS: digits are transmitted at the start of the call, with ANI the digits are the number of the originating telephone line. These digits can be used by a VRU to block a call, route it to the responsible agent or do a database lookup to retrieve the caller's account information. There are several standards for the way phone companies deliver ANI digits. Local phone companies, like New York Telephone, deliver ANI digits in between the first and second ring. This service is known as caller ID. They sell it as a service for their residential subscribers to ward off harassing calls, but caller ID also has business applications.

A more useful ANI service is delivered by the long distance companies, such as AT&T, MCI, and Sprint, through their 800 lines. If you have an 800 number and you have it directly hooked up to your office on a T-1 trunk then you can receive the phone number on all the 800 calls you receive. This service is now so widespread that, with an 800 line from virtually any long distance company (not just the three I mentioned above), you can receive the phone numbers of over 95% of all the phones in the U.S. The ones you can not get are typically in outlying rural areas. How you will receive ANI digits on your 800 lines depends on your chosen long distance carrier. There are basically two ways to receive ANI from a long distance carrier C in-band or out-of-band. In-band means you receive the digits in the voice data link of the conversation. Out-of-band means you receive the ANI digits on another circuit, typically in a communications channel just devoted to signaling. At the time of writing, AT&T, for example, only provides ANI on PRI ISDN trunks using out-of-band signaling. ISDN stands for Integrated Services Digital Network. PRI stands for Primary Rate Interface. PRI ISDN trunks are an extension of today's common T-1 digital standard. But they are not the same as today's T-1.

MCI will provide ANI services in-band and out-of-band, on T-1 or directly on an individual 800 line. When you buy several 800 lines, it's less expensive to connect from your office to your chosen long distance carrier on a T-1 line.

This chapter was adapted from Bob Edgar's book: PC Telephony, available from Flatiron Press, 800-LIBRARY. I thank Bob, founder of Parity Software Development Corp for his help and advice with this chapter.

Section III

Understanding Facsimile Technology

Chapter 6

The Fax Image

Resolution

A fax image is made up of dots. The dots cover the page and when combined, compose letters, numbers, pictures, etc. There are two different resolutions widely used by fax machines and fax boards. These resolutions are referred to as standard and fine. Most fax machines have some way of switching between the two resolutions. The default is usually standard.

Standard resolution is 204 x 98 dpi or 204 dots per inch in the X (horizontal) direction and 98 dots per inch in the Y (vertical) direction. The dots are not square, but rectangular. Fine mode doubles the Y resolution of standard mode; thus it is 204 x 192 dpi. Sometimes the numbers are rounded off, so people may refer to a standard mode fax to be at 200 x 100 dpi and fine mode to be 200 x 200. This gives a good comparison to other devices that you may be familiar with such as ink jet printers and laser printers of several years ago which had 300 x 300 dpi resolution, or the newer laser printers which have 600 x 600 dpi resolution. The more dpi, the clearer the image on the paper. However, the more dpi, the bigger the file, which is very important when considering fax communications since this file needs to be sent across telephone lines. The bigger the file, the longer it takes to send the image. Fine mode images take roughly twice as long to send as standard mode images.

Compression

When facsimile systems first became popular, with the establishment of Group 1 and Group 2 standards, they handled information in real time as an analog signal, in which variations in the voltage on the line were an analog representation of the intensity of the signal. The idea of processing the signal by clocking, quantizing, and storing it as digital information seemed like something that would be found aboard the Starship Enterprise and not on an office desk. In those old analog days scanning and recording systems were mechanical. It was difficult to even mark a straight line down a received page without noticeable jitter (a sawtoothed effect in the printout).

Keeping the scanning and recording spots in exact step with the clock was impossible. There was also a drift between the clocks at the transmitter and receiver units. Although manual adjustment of the frequency kept the skew within acceptable limits, this was far from being an ideal solution. There existed no digital modem for sending signals and providing a locked-in sampling clock at the receiver. Practical means for the digital storage of pil information for one or two scanning lines were yet to be developed.

As digital technologies evolved, it finally became practical to use compression encoding techniques to remove redundancy (repetitious information such as white space), from the page being scanned and restore pil information for recording at the receiving end.

Compression is defined as the ability to change the form data into a representation expressed by a minimum of bits, reducing transmission time. Encoding schemes reduce sending errors and further speed broadcast times. This technique has been viewed as a leading factor in the growth of facsimile technology, transforming it from a curiosity of limited use into a ubiquitous utility, very much like indoor plumbing and electricity.

For example: If a piece of paper is 8.5 inches wide by 11 inches long and each square inch contains roughly 200 X 100 dots (in standard mode), then there are 8.5 X 11 X 200 x 100 dots for a single page. That's a lot of dots. 1.87 million to be precise. If we were to represent each dot as a bit, then a single page would take up 1.87 million /8 or 234 Kbytes (8 bits to a byte). At normal fax speeds of 9600 bits per second, a file this big would take about 6 minutes to send (trust me on the math). But most of know that a single page does not take this long to fax. How is this done?

Rather than send a bit for each potential dot on the page, the file is compressed by taking advantage of the fact that most fax pages have big chunks of space where the color does not change. The standards committees that govern fax communications used a method for compressing such data that is called Modified Huffman, sometimes referred to as MH. It is also referred to as Group 3 compression (after those fax machine that use it) and one-dimensional compression, which refers to the fact that the compression deals with each line of dots one at a time and uses no knowledge of previous lines to build subsequent lines. Another name is T.4 compression. T.4 is the name of one of the international standards specifications that defines formats for images. Modified Huffman compression will typically shrink the size of an image down to 20 to 30 Kbytes per page. Thus the time to send one page is about 20 to 30 seconds, more in line with what you are probably used to if you have watching fax machines send or receive faxes.

Note that this example is a standard mode example. The same process occurs on fine mode faxes, but they will still be about twice as large as the standard mode faxes.

After the standards committees agreed on the Modified Huffman standard, they later agreed on two other compression types that are similar in concept to Modified Huffman but use a different algorithm and typically compress images even more densely than Modified Huffman. The first is called Modified READ (MR) or Group 3, 2-D (for 2 dimensional). The difference in size between an MH compressed file and an MR compression file is roughly 15%. This percentage varies depending on the composition of the image. The other compression type is Modified Modified READ (MMR), also known as Group 4 compression (since it is the compression type specified for Group 4 fax machines). The standards committees refer to the T.6 specification for the definition of this image type. An MMR image is roughly 25% to 30% more dense than an MH image.

Not all fax machines and boards support the advanced compressions (MR & MMR), but all support MH. Most machines support MR, and about 20% support MMR. Generally, only the intelligent fax boards support either MR or MMR. The amount of money that can be saved on one's telephone bill by using the advanced compressions can be significant if there is a lot of fax traffic that is going long distance or internationally. Since the files may be up to 30% smaller, the telephone bill can be reduced by up to 30%.

How Compression Occurs

In its Recommendation T.4, the ITU-T specifies that Group 3 fax units must incorporate MH run-length encoding of scanning lines. The technique used is called run length encoding. The algorithm begins scanning at the upper left-hand corner of the image, and the image is surveyed at a specific resolution value, typically 203 dpi, a line at a time, looking for consecutive dots; that is, how many of these dots (or pils) of black or white there are before there is a transition to the opposite color. All facsimile encoding techniques operate by scanning one line at a time and then going on to the next. This is best visualized by considering the lines across a pad of notebook paper. These would be the scan lines the algorithm processes. Scanning takes place from left to right. The survey algorithm begins by determining the color of the first pil (black or white), and then counts how many dots are that color: 5, 23, or maybe the entire line, 1728 bits.

Consequently, there is no direct storage of black and white pils, only the run lengths themselves are stored. For example, the compression algorithm may determine that there are 23 white pils. This information is not transmitted as the 23 zero bits or even as a byte representing the number 23, but rather as a much shorter encoded version—a code word—of that number: 11 bits that represent the number 23. These code words have been standardized and embodied into tables, and are well-established. Thus, instead of sending a white line across a page as 1728 bits, MH sends a 9-bit code word. This compresses the information 192 times. Different code words are used to represent shorter white (or black) runs.

A total of 1728 code words would be needed to cover all possible run lengths across a page. To produce a shorter code table, the runs were grouped in multiples of 64 words in a make-up code table, with shorter runs of 0 to 63 placed in a separate, terminating code table. This reduced the matter down to only 92 binary codes for any white run lengths of 0 to 1728 pils by sending a make-up code followed by a terminating code. Black runs also have 92 different codes. These code words were chosen by looking at varied page samples and minimizing the numbers of bits needed to send each one.

Compression Techniques

Although all compression techniques operate by surveying one scan line, then proceeding on to the next, there are important differences between them.

Modified Huffman compresses only one scan line at a time; it looks at each scan line as if it were the only one. Each line is viewed by it as a separate, unique event, without remembering anything about previous scan lines. Because of this single-line characteristic, MH is referred to as a one-dimensional compression technique.

Modified READ on the other hand, uses the previous line as a reference, since most information on a page has a high degree of vertical correlation "vertical correlation"; in other words, an image (whether it is a letter or an illustration) has a continuity up and down, as well as from side to side, which can be used as a reference. This allows MR to work with only the differences—the variable increments (rates of change), or deltas—between one line and the next. This results in a rate of about 35 percent higher compression than is possible with MH. MR operation may at first appear as a complex concept to grasp, but it really is not. Imagine, for example, a white sheet of paper with a black circle in its middle. The sheet, being all white can be effectively compressed because, after the first compression, there is no change in the following lines so the MR algorithm just repeats the line.

When MR starts scanning the circle, black run lengths begin and, as it proceeds downward, the circle scan gets increasingly larger. However, since MR uses previous lines as a reference, instead of counting all the black pils to encode them as the MH technique would, the MR algorithm only notes the deltas, or rate of change, between lines. Therefore, the difference between MH and MR is, essentially, that the latter uses a knowledge of previous lines to reference its vertical compression. Because MR works vertically as well as horizontally, it is called a two-dimensional compression technique.

As the name implies, MMR is very similar to MR. The difference lies in how it handles errors.

When a fax signal is distorted by a noise pulse induced by electrical interference, or whatever reason, errors occur in bits transmitted over the phone line making impossible the recovery of the original pattern of black and white pils in the received copy. To prevent an incorrect pattern from propagating down the page and corrupting the entire message, a form of error correction mode (ECM) was needed.

Originally, facsimile technology had no protocol for a recovery method in cases when bits or code words were lost or garbled during transmission over the telephone network. The original Group 3 compression techniques did not have an ECM, and therefore a means was arrived at through which to resynchronize the printer when an error took place; otherwise, the entire fax could be lost. This is why both MH and MR contain a special signal called an end of line (EOL) code. If the printer receiving the fax runs into an error, it waits until receiving an EOL code in the stream, to restart its decompression procedure and resynchronize. That way, if data errors occur, the entire message is not lost. Often, rather than producing a white line, the printer will simply repeat the previous line. Generally this is not immediately obvious to the naked eye, particularly in text transmissions. In any case, although the fax may not end up looking particularly attractive, only a few scan lines have been lost, not the whole thing.

On the downside, the EOL code adds an additional 12 to 24 bits of data at the end of each scan line, lengthening transmission time.

One of the differences between MR and Modified Modified READ, is that MMR does not use the EOL code because it was designed originally for Group 4 facsimile work. Group 4 only operates on a digital network, which is inherently error-recovered because if an error occurs, the data packets are called for again and repeated. Since there is no data loss, there is no need to resynchronize the image and this results in a 12- to 24-bit-per-scan-line compression gain. Therefore, statistically, MMR provides the highest compression for normal images.

One of the important facets of compression techniques is how often a reference—versus a delta—line is sent. This is known as the K factor. With standard mode fax under MR, K is equal to two (K=2); that is, every second scan line is a reference line and then a delta line is sent. At fine resolution K=4, which means that every fourth line is a reference line. MMR's K factor equals infinity because MMR uses no MH coded lines. MMR sends a reference line at the beginning of a page, then deltas all the way down. An error-free signal is needed for this to work. If this were done in MR, without error checking, an error on any one line would propagate downward, worsening with every additional line, corrupting the information on the balance of the page.

In 1992, the Group 3 Study Group of the ITU-T allowed Group 4 (T6 encoding) compression—a variation of MMR—to become an optional transmission technique for Group 3. Since MMR can only be transmitted by a Group 3 device supporting "ECM", both the transmitting and receiving units must support ECM to work with that type of data stream.

It is interesting to note that there are times when any of these compression techniques can go the other way and generate a data stream considerably larger than the actual pil-per-pil representation. This sometimes occurs during the processing of some grey scale images (such as photographs, or desktop publishing generated documents where grey is used) where there are numerous changes, black and white, horizontally and vertically, requiring the transmission of many additional code words. This is why computer-generated grey is very slow to transmit.

Compression, Resolution, and Speed

For all intents and purposes, compression and resolution are unrelated. MH, MR, and MMR are strictly loss-less compression techniques—the compressor and decompressor algorithms that run are solely concerned with the number of pils to survey in any given dimension—therefore none of these techniques has any effect whatsoever on resolution.

This is also why speed is not a factor in compression. Compression techniques are applied to bit-map images and are unaffected by scanner or modem speeds. The evidence for this is that it is possible to have a bit-map on a computer's hard disk, apply compression techniques to it, and obtain a compressed image, all of it without the involvement of transmission or facsimile technologies.

Anyone attempting to estimate and plan for a company's fax expenses needs to understand that since different compression techniques result in different file sizes and storage requirements. An efficiently compressed file is smaller and requires less transmission time. For a user doing multipage broadcasts products that support MMR, the highest compressed image, through optional Group 3 parameters, can be a matter of vital importance to the bottom line.

The use of faster modems, transmitting at 14.4 kbps, offers the possibility of sending more bits per unit of time over the telephone. But this is solely a matter of broadcasting data faster across the wire—the level of compression (whether MH, MR, or MMR are used) remains the same. This is purely a matter of raw transmission speed, completely unrelated to image, scanner, or compression. If we think of the compressed data as a letter, then it becomes obvious that it could be sent either by train (slow modem) or plane (fast modem). Although the trip would take less by plane, whether sent by land or air it would still be the same letter. A transmission rate is not dependent on the sending machine, but on the receiving machine.

A traditional fax machine is a mechanical device. It must reset its scanner, and advance the page as it prints each scan line it receives. Today's machines generally have a 10-millisecond scanning line time requirement.

When fax devices warble lovingly at each other before transmission, during the handshake period, they exchange information on scanning capabilities. If the receiving machine's rate is 20 ms per line and the transmitting machine sends data at a faster rate, it will add fill bits (also called zero fill). These extra bits pad out the amount of send time, giving the remote machine the additional time it needs to reset prior to receiving the next scan line.

The amount of fill bits added by the sending machine is determined by the receiving machine's capability. If a machine that sends at high speeds broadcasts to a slower machine, some of the time-saving benefits of sending encoded data and higher modem speed may be lost if a lot of fill must be introduced.

The rate at which a machine can receive is not related to the rate at which it can send. Although CBF generally does not require fill bits, if a CBF device is sending to a traditional fax machine it must tailor its sending mode to that device's receiving capabilities.

It is important to bear in mind that because facsimile is a compressed technique, in the case of CBF, it is essential to have software-based tools that will enable the user to correctly decompress the image back into an accurate, readable form. CBF requires a knowledge of compression techniques and their compression ratios, that is, the number of bytes it takes to store images, in order to accurately estimate archiving capacity. File sizes obviously are a factor in how long a telephone call will take—the latter being directly related to cost.

Getting Hard Copy

When the receiver modem decodes the received analog fax signal regenerates the digital signal sent by the fax transmitter, the MH/MR/MMR block then expands this fax data to black-white pil information for printing.

Overall, there are two ways to convert a fax into hard copy: through a thermal printer, or a regular printer (the latter covering everything from pin to the preferred laser printer formats).

Each inch of a thermal printer's print head is equipped with 203 wires touching the temperature-sensitive recording paper. Heat is generated in a small high-resistance spot on each wire when high current for black marking is passed through it. To mark a black spot, the wire heats from non-marking temperature (white) to marking temperature (black) and back, in milliseconds.

Since thermal paper comes in rolls, a thermal printout has the advantage of adapting itself to the length of the copy being transmitted: if the person on the other end is sending a spreadsheet, the printout will be the same size as the original. With sheet-fed printers, such as a laser printer, the copy is truncated over as many separate sheets as necessary to download the information. That, however, is about the only major advantage of thermal printouts. In general, thermal printouts offer less definition, thermal paper is inconvenient to handle due to its tendency to curl, and if the copy is to be preserved, it must be photocopied because thermal paper printouts tend to fade over time. Then, most important of all, there is the matter of cost: thermal paper printouts are between five to six times more expensive than those made on plain paper. Plain paper fax machines using laser printer technology are becoming increasingly popular, particularly in the U.S.

TIFF and PCX

When a fax machine sends or receives an image, it goes from paper to paper, being converted into an image as described above. When a computer sends or receives a fax, it stores the image as a file or series of files. Two of the common file formats for storing fax images are TIFF (Tag Image File Format) and PCX (PC Paintbrush format). Both of these formats are widely used by other computer programs, such as drawing programs, graphics programs, scanning programs, etc. The fact that standard formats are used gives the user a lot of flexibility in manipulating fax images, combining them with other images, adding them into word processing documents, etc.

TIFF files of the same image typically take up about half the space of PCX files. So if you plan on setting up a system that will either contain a lot of files or receive a lot of files, you might want to remember that systems that support TIFF will require about half the disk space of systems that use PCX. Average image sizes will range from 20 - 40K per page for TIFF files and 40 - 80K per page for PCX files. The amount of space required is dependent primarily on the resolution of the page and the composition of the image. Pages containing text will usually be much smaller than pages containing pictures or drawings with large areas of gray.

Chapter 7

The Complete Fax Call

T.30

Dial to Acknowledgement

What takes place before, during and after a fax transmission may seem to fall into the realm of magic, but is in reality a very carefully orchestrated procedure. With small variations, a modern fax machine call is made up of five definite stages as defined by the ITU-T's T.30 specification: Phase A, the call establishment; Phase B, the pre-message procedure; Phases C1 to C2, consisting of in-message procedure and message transmission; Phase D, the post-message procedure; and Phase E, call release.

Phase A

Phase A takes place when the transmitting and receiving units connect over the telephone line, recognizing one another as fax machines. It is here that the handshaking procedure begins between the transmitting and receiving units. This is accomplished through a piercing, 1100-Hz tone (the calling tone or CNG), sent by the caller machine. Virtually everybody is well-acquainted with this mating call, having heard it either by dialing a fax number by mistake, or waiting to hear the answering love warble of the machine at the other end before activating your own to transmit the message.

It is here that the caller machine sends its Called Station Identification (CED) and the answerer replies with its own, a shrill 2100-Hz tone. Once this has been accomplished, both machines move on to the next step.

Phase B

In Phase B, the pre-message procedure, the answering machine identifies itself, describing its capabilities in a burst of digital information packed in frames conforming to the High-Level Data-Link Control (HDLC) standard. Always present is a Digital Identification Signal (DIS) frame describing the machine's standard TSS features and perhaps two others: a Non-Standard Facilities (NSF) frame which informs the caller machine about its vendor-specific features, and a Called Subscriber Identification (CSI) frame. At this point, optional Group 2 and Group 1 signals are also emitted, just in case the machine at the other end is an older Group 1 or Group 2 device unable to recognize digital signals. This is very rare, however, and chances are that both devices will be Group 3 machines, in which case they will proceed from the identification section of Phase B to the command section.

In the command section of Phase B, the caller responds to the answerer's initial burst of information, with information about itself. Previous to transmitting, the caller will send a Digital Command Signal (DCS) informing the answerer how to receive the fax by giving information on modem speed, image width, image encoding, and page length. It may also send a Transmitter Subscriber Information (TSI) frame with its phone number and a Non-Standard facilities Setup (NSS) command in response to an NSF frame. Then the sender activates its modem which, depending on line quality and capabilities of the machines may use either V.27 (PSK) modulation to send at a rate of 2400 or 4800 bps, or V.29 (QAM) modulation to transmit at 7200 or 9600 bps.

A series of signals known as a training sequence is sent to let the receiver adjust to line conditions, followed by a Training Check Frame (TCF). If the receiver receives the TCF successfully, it uses a V.21 modem signal to send a Confirmation to Receive (CFR) frame; otherwise it sends a failure-to-train signal (FTT) and the sender replies with a new DCS frame requesting a lower transmission rate.

Phase C

Phase C is the fax transmission portion of the operation. This step consists of two parts: C1 and C2, which take place simultaneously. Phase C1 deals with synchronization, line monitoring, and problem detection. Phase C2 includes data transmission.

Since a receiver may range from a slow thermal printing unit needing a minimum amount of time to advance the paper, to a computer capable of receiving the data stream as fast as it can be transmitted, the data is paced according to the receiver's processing capabilities. An Error Correction Mode (ECM) procedure encapsulates data within HDLC frames, providing the receiver with the capability to check for, and request the retransmission of garbled data. Because of its encapsulation within the HDLC frames this on-going procedure does not lengthen transmission time.

Phase D

Once a page has been transmitted, Phase D begins. Both the sender and receiver revert to using HDLC packets as during phase B. If the sender has further pages to transmit, it sends a frame called the Multi-Page Signal (MPS) and the receiver answers with a Message Confirmation Frame (MCF) and Phase C begins all over again for the following page. After the last page is sent, the sender transmits either an End Of Message (EOM) frame to indicate there is nothing further to send, or an End Of Procedure (EOP) frame to show it is ready to end the call, and waits for confirmation from the receiver.

Phase E

Once the call is done, Phase E, the call release part, begins. The side that transmitted last sends a Disconnect (DCN) frame and hangs up without awaiting a response.

Chapter 8

Facsimile Standards

Introduction

As we've seen, the beginning of facsimile technology is traceable to the 1840's, when the first successful fax device was patented; and, commercial fax service began in France, in 1865.

After the Second World War there was great interest on the part of publishers in using fax technology to send newspapers directly to subscribers homes, but the coming of television as well as technical problems forced an abandonment of the idea.

Over the 1950's and 1960's, fax technology began to evolve into the form we recognize today. As fax became more widely used, it soon became obvious that standards would be needed to enable different fax machines to communicate with each other.

It was not until October 1966 that the Electronic Industries Association proclaimed the first fax standard: EIA Standard RS-328, Message Facsimile Equipment for Operation on Switched Voice Facilities Using Data Communication Equipment. This Group 1 standard as it later became known, made possible the more generalized business use of fax.

Although Group 1 provided compatibility between fax units outside North America, those within still could not communicate with other manufacturers units or with Group 1 machines. Transmission was analog, typically it took between four to six minutes to transmit a page, and resolution was very poor.

U.S. manufacturers continued making improvements in resolution and speed, touting the three-minute fax. However, the major manufacturers still used different modulation schemes FM and AM. So, again, there were no standards.

Later in 1978, the ITU-T (then referred to as the ITU-TSS) came out with its Group 2 recommendation, which was unanimously adopted by all companies. Fax had now achieved worldwide compatibility and this, in turn, led to a more generalized use of fax machines by business and government, leading to a lowering in the price of these units.

When the Group 3 standard made its appearance in 1980, fax started well on its way to becoming the everyday tool it is now. This digital fax standard opened the door to reliable high-speed transmission over ordinary telephone lines. Coupled to the drop in the price of modems, an essential component of fax machines the Group 3 standard made possible today's reasonably priced, familiar desk top unit.

The advantages of Group 3 are many; however, the one that quickly comes to mind is its flexibility, which has stimulated competition among manufacturers by allowing them to offer different features on their machines and still conform to the standard. The improvement in resolution has also been a factor. The standard resolution of 203 lines per inch horizontally and 98 lines per inch vertically produces very acceptable copy for most purposes. The optional fine vertical resolution of 196 lines per inch improves the readability of smaller text or complex graphic material.

Group 3 fax machines are faster. After an initial 15-second handshake that is not repeated, they can send an average page of text in 30 seconds or less.

Group 4 emerged as a standard in the early 90's. It was designed to greatly speed up fax transmission, to allow faxes to be addressed to individual recipients (also called subaddressing) and to allow future improvements such as color fax. Technically it was a great leap forward; however, it never became a reality because it required a high speed digital ISDN line. Since ISDN was slow to become accepted, especially in the United States, companies did not bother with the expense of Group 4 because they knew there was only a very small population of Group 4 machines they could send to.

Within a few years it became apparent that Group 4 wasn't going to gain wide acceptance, so the standards committees went back to Group 3 and added much needed features to it. One of the features was sub-addressing, which was standardized in the fax protocol. Computer fax vendors who wanted some way to electronically address faxes (beyond just the phone number) had lobbied hard for this feature. Unfortunately, the sub-addressing standard came too late to become accepted by fax users. By this time millions of fax machines were in use and users generally placed the paper in the tray and dialed the fax number. Retraining them to also enter a recipient's sub-address was too much to ask. Thus the long awaited T.30 sub-address feature became part of the Group 3 standard, but everyone in the fax industry generally ignores it.

The useful enhancements to the fax standards have been in speed of transmission. Although many fax machines still use 9600 bps, 14.4 kbps (V.17) fax speed is now widely accepted, and currently work is being done on the V.34

standard which would bring the speed up to 28.8 kbps.V.34 also speeds up the handshake time for setting up and breaking down a fax call between two fax devices. Currently it can take 15 to 20 seconds to connect a call before starting to transmit, and another 6 seconds at the end of a call.The V.34 standard can reduce the time to 8 seconds for setup and 1 second for breakdown, a time reduction that can result in significant cost savings when enough of the V.34 fax devices are in use.

IP Fax Standards

As IP fax becomes more generally deployed, standards are becoming important to provide interoperability between systems. IP fax standards work has been progressing steadily for a number of years now.

Store-and-forward fax - The IETF (Internet Engineering Task Force) has been working on an addendum to the SMTP (Simple Mail Transport Protocol) which involves a MIME attachment for fax. This attachment uses the TIFF-F format for normal fax and TIFF-FX for more advanced fax features. The SMTP standard integrates fax with email.

Real-time fax - These standards are being developed by the ITU' s TR-29 Committee. The ITU and IETF are working together to ensure that these two standards don't conflict or duplicate each other.

For the latest information on fax standards visit:
www.itu.int and www.ietf.org and www.imc.org/ietf-fax

Section IV

Computer Based Facsimile

Chapter 9

Fax Routing

Introduction

The Group 3 protocol (the digital fax standard that determines a fax machine's capabilities) does not stipulate any kind of an auxiliary address for the inward routing of faxes. It was not until the appearance of more complex systems such as fax machines designed to keep faxes internally, in secure or private mailboxes, that a routing mechanism of sorts first became necessary. With the coming of CBF and its inherent flexibility, the possibilities opened by various routing schemes became too valuable to ignore.

The widespread use of LANs has made dependable inward routing (the process of selecting the correct destination for a message within an organization), a necessity. There are now several ways to route messages. Which one is selected and how it is implemented, has a direct effect on network fax capabilities. When studying routing schemes, a prime consideration must be the installed base of Group 3 fax machines, estimated at 50 million. If they cannot readily use the routing solution, it has missed the mark. The next most important concern must be the outside sender.

A basic consideration in routing system selection is what the sender of the fax must do for it to work. The different ways people send faxes preclude the wide use of some of the more elaborate sender-dependent routing mechanisms.

Methods requiring a trained sender generally serve a limited purpose. If a company needs to have an inward routing method used by a closed user community within its organization for a specialized task, then it can train personnel, use specialized fax machines, and create something unusual in the way of a routing scheme. The solution then meets a special need while posing no problems because the company controls the community of senders and receivers whose efficiency is increased by the system. However, in more widespread applications where the volume of faxes received is in the tens, hundreds, or even thousands of faxes a day, coming from all kinds of machines and senders, another solution is advisable.

Manual Routing

In its most basic form, manual routing consists of someone picking up incoming faxes from a regular fax machine and distributing them to the intended recipients. With CBF, manual routing uses a software utility to electronically display to an operator a listing of faxes received. Usually the information displayed comes from the cover page as well as the transaction record that shows the date and time the fax was received, how many pages it has, and how long it took to receive. The operator doing the routing scans this and can probably determine from this information whom the fax is intended for. Sometimes, however, this is not enough and the fax itself must be viewed before routing; this may be unsatisfactory when dealing with proprietary or confidential information.

DTMF Routing

Unattended DTMF routing uses the buttons on a TouchTone phone. A concern with DTMF routing is that it depends on the sender to work: anyone attempting to send in a fax must be trained in the particular routing system or somehow get operating instructions.

With DTMF routing the sender not only needs a telephone number, but an extension number as well. This may pose a problem. How is that additional set of numbers used? When the sender dials, he may expect an operator to come on. If there is no operator, suddenly there is a tone. Is the extension number entered now? What happens if a digit is missed? Then, if the sender takes too long to decide, the machine on the other end may time out, forcing the caller to start all over again.

Often, when the sender hears a sound different from the regular dial tone, the tendency is to punch the start button on the fax machine. Automated voice prompting helps unless dealing with a fax machine at the sending end instead of a human ready to enter additional number sequences when told to. Generally, if a second sequence of numbers is not entered, the fax is diverted to a default central directory for subsequent manual routing.

Direct Inward Dialing

DID is generally viewed as the best routing choice. This is because, unlike DTMF, it is the telephone company's central office (CO) not the end user that provides the routing information. All the sender does is dial a single telephone number, and the fax goes directly to the recipient's workstation. DID is the most foolproof, transparent way of routing.

DID consists of one or more trunk lines between the CO and a customer's premises. DID trunks are normally (but not in CBF's case) used to reduce the number of channels between the PBX and the CO: they are one-way trunks. A PBX perceives the DID trunk as one of its single-line phones and can interpret four-digit dialing. Therefore, some switching functions usually done at the telephone company's CO are done at the customer's premises. Generally, each CBF board can have only one DID line, but an unlimited number of phone numbers can be installed on that line since the board hardware and software direct the incoming call from the CO.

When outside callers dial a station on a DID-configured PBX, the CO switch signals the PBX for service. An available trunk is activated by the CO by causing a current to flow. The PBX detects the current flow but, rather than responding with a dial tone (which it would have done if it really considered this loop to be a true station-to-PBX loop), it gives the CO a wink-start signal. The wink informs the CO that the PBX is ready to receive the incoming addressing information. The wink signals the CO switch to send the identifying digits dialed by the distant user, which the PBX uses to complete the connection. These last digits provide the data that the system uses for inward routing. After the pulses have been received by the PBX and the desired station has answered the call, the PBX signals the CO that the call and billing can begin, by reversing the polarity of the current as it did for the wink-start signal, but leaving it reversed for the entire duration of the call. When the call is finally disconnected, the DID facility returns to its idle state.

DID installation requires an awareness of polarity and power requirements. With standard conventional telephone facilities this is not a consideration, but with DID it is all-important. Any business considering the use of DID should use its in-house resources and invite the participation of its MIS, computer, and telecommunications experts in planning for the company's network fax system. Once the DID system is up and working, no further attention is needed and it will operate transparently for both sender and recipient.

DID also requires the user to maintain line voltage. In the familiar telephone services, the CO maintains the 48 volts necessary for the line to operate, with battery back-ups in case of a blackout. With DID, since the user provides the power source, if line voltage is lost, the telephone company shuts down the line and callers get a busy signal until it is reactivated by the telephone company. At this point in the planning stage, the advice of the local telephone company technician on how to prevent your DID installation from going down will prove useful.

OCR Routing

When the problem of handwriting recognition is solved, optical character recognition (OCR) technology undoubtedly will play a major part in routing. Meanwhile, OCR routing depends on conventions such as code numbers appearing within a specific area of the cover page, combinations of forward and backward slashes (//\\), or bar codes similar to those in use by grocery stores. These options require senders to be trained or use specific formats, at least for the cover page. There are OCR software utilities that can scan any kind of cover page and attempt to find a name, for instance, by attempting to recognize what is written after TO: by mapping it through a stored name pattern. Accuracy is about 80 to 90 percent for a typed document; if unsuccessful, the system defaults to some form of manual routing.

Chapter 10

Application Programming Interfaces (APIs)

Introduction

An Application Programming Interface (API) is software that an application program uses to request and carry out lower-level services performed by the computer's or a telephone system's operating system. For example, for Windows, the API also helps applications manage windows, menus, icons, and other graphical user interface (GUI) elements. In short, an API is a "hook" into software. An API is a set of standard software interrupts, calls and data formats that application programs use to initiate contact with network services, mainframe communications programs, telephone equipment of program-to-program communications. For further example, Standardization of APIs at various layers of a communications protocol stack provides a uniform way to write communications. NetBIOS is an early example of a network API. Applications use APIs to call services that transport data across a network.

There are thousands of APIs. The most relevent to Computer-Based Fax Processing are described briefly.

Telephony Applications Programming Interface (TAPI)

We can easily imagine an office in the not too distant future where different messaging media: voice, text, graphics images and video are integrated and controlled from the desktop PC. To make a phone call, you would pull up a phone book on the screen, highlight the person or company you want to call, and click on the "Call" button with your mouse. To make a conference call with two of your colleagues, you would pull up the company directory and drag the parties' names into the "Conference" box. To check for messages, you would open your on-screen "in-box" to see a list of new and saved voice, text, image and video messages. To reply, you would compose a document in any application, and drag the document to the "Reply" button. The PC might determine the medium and means of communication required to deliver the reply.

Devices such as PBXs, fax machines, desktop telephones and printers have traditionally been built to accept and/or transmit information on their specialized medium, ie. a telephone line or a sheet of paper, but not to provide intelligent, two-way communication with a controlling desktop computer or network. A printer, for example, can only transmit a few rudimentary messages back to a computer: "out-of-paper", or "on/off-line". This lack of communication is despite the presence in most of these devices of sophisticated microprocessors comparable in power at least with early PCs.

In the future, printers will have greatly enhanced abilities to communicate: from your PC, you will be able to see the number of sheets of paper in the feeder, check the level of toner in each color cartridge, send a command to switch to stationery instead of blank paper, and so on. The desktop telephone will have an RS-232 serial port or other connection to a computer which will enable two-way communication: the computer will be able to ask the telephone to dial a number, for example, and the telephone will send status indications back to the PC similar to those displayed on the phone through colored lights or an alphanumeric LED panel. Ultimately, the telephone itself may become an expansion card with a socket for a handset or headset with microphone and loudspeaker.

The company phone system will be controlled, probably via a serial link, by a PC attached to the corporate LAN instead of relying entirely on its internal CPU. When an incoming call is signaled, the PBX will notify the computer, which will reply with a command telling the PBX where to route the call. (Such PBXs, so-called "dumb switches," are already available from companies such as Summa Four). To make a conference call, the controlling computer will send a command to the PBX. The voice mail PC will be connected to the PBX as a series of extensions and also to the controlling PC via a LAN.

Desktop Applications

Microsoft has a clear vision of how this technology should be tied together: via graphical applications running in a Windows environment. As a first step, Microsoft, in conjunction with Intel, drafted version 1.0 of the Windows Telephony Applications Programming Interface, or "TAPI", which was published in May 1993. This is an attempt to define a set of function calls (the API) which will be used by Windows programmers to interact with telephony devices such as telephones, telephone lines and telephone switches. TAPI is sometimes referred to simply as "Windows Telephony."

Some important categories of application which will incorporated TAPI include:

Integrated Messaging

A single system for receiving, archiving and responding to voice, text, graphics and video messages.

Personal Information Managers

These will include facilities for automated dialing and collaborative computing over telephone lines.

Advanced Call Managers

Do you really know how to use the conferencing feature on your company telephone system? How to ask for "camp on", where a call will be put through to your colleague who's currently on the phone as soon as he hangs up? How to put that "At lunch" message on the phone display panel to notify callers to your extension that you're away from your desk? I don't know how to use half the features on our digital phone system, and I know I'm not alone. I'm looking forward to an application that gives me access to the features of our phone system through graphical menus on my PC. Through advanced call managers, Microsoft is envisaging control of your phone system through Windows.

Microsoft has since published several later TAPI specifications. For recent information, visit: www.microsoft.com

Communicating Applications Specification (CAS)

First introduced in 1988, CAS is a high-level API developed by Intel and DCA to define a standard software API for fax modems. CAS enables software developers to integrate fax capability and other communications functions into their applications. The next version, CAS 2.0 was development by Instant Information Inc (I3), which in 1997 was acquired by FaxBack.

Software developers worldwide use the FaxBack CAS API due to its simple and flexible architecture suitable for custom fax applications that include fax servers, fax broadcast systems and fax-on-demand services.

CAS' unique architecture offers flexibility for easily passing data between servers and clients. With CAS, clients and servers exchange all data in the forms of lists of information elements called "tags." Using CAS, organizations can develop manageable, scalable fax applications rapidly to support the most demanding fax communications requirements.

An important advantage of the CAS interface is the ability to combine ASCII text files with graphics files in PCX and DCX format in creating fax documents. The CAS manager will also automatically construct transmission cover pages, including date, time, sender and message text fields. Another useful feature is the ability to perform a high-speed, error-correcting file transfer to another device supporting CAS.

The CAS programming model deals with events. An event is a single phone call involving the fax board and a remote device such as a fax machine or another fax board. An event is one of the following types:

Send The computer makes a call and initiates a transmission of one or more files to a remote device (fax machine or fax board).

Receive A remote device makes a call and initiates a transmission of 1 or more files to the computer. A program can obtain information about receive events which have taken place by querying the Receive queue.

Polled Send The computer waits for a remote device to call and then starts sending information to it.

Polled Receive The computer makes a call to a remote device, then receives a transmission from it.

Each event is assigned an "event handle" by the CAS manager for the given board.The event handle will be a number in the range 1 .. 32767, which uniquely identifies the event. No two events for the same board will have the same handle.

Each event has an associated "Event Control File" which contains information about the event, such as the date, time, phone number, file name(s) etc. For events initiated by the computer (Send and Polled Receive), Event Control Files are created by the functions making the request. For events initiated by the remote device

(Receive), the Event Control File is created by the CAS manager.

A File Transfer Record is included in the Event Control File for each file transfer operation associated with the event.

An Event Control File created by a Receive event will contain one File Transfer Record for each file received, the CAS manager is responsible for this process.

There are three queues of events maintained as linked lists of Event Control Files by the CAS manager for each installed board:

Task Queue The Task Queue contains an Event Control File for each event which has been scheduled, but which has not yet been executed.

Receive Queue The Receive Queue contains an Event Control File for each completed Receive or Polled Receive event.

Log Queue The Log Queue contains an Event Control File for each event which has been successfully or unsuccessfully completed.

Queues are maintained by the CAS manager sorted in chronological (date and time) order.

An event retains the same event handle even when it is

moved from one queue to another. For example, a Send event will be moved from the Task Queue to the Log Queue when it completes, but will still be identified by the same event handle.

A completed Polled Receive or Receive event will appear in both the Receive Queue and in the Log Queue. Two copies of the Event Control File exist in this case C if the event is deleted from one queue, it will still appear in the other.

Other events will appear in one queue only:

- Completed Receive and Polled Receive events: Receive and Log Queues.
- Pending Send and Polled Receive events:Task Queue.
- Completed Send events: Log Queue.

If an event is in the process of being executed, it is known as the Current Event, the Event Control File for the Current Event is not in any queue.

The FaxBack CAS 2.0 Software Development Kit (SDK), first available in September 1997, provides developers with all of the necessary tools needed to build custom fax applications.

For more information, visit: www.faxback.com

Chapter 11

Application Generators

Parity Software's Graphical VOS Programming Language

An Application Generator, (also referred to as an App Gen) is software that writes software. Application generators are software tools that, in response to your input, write software code a computer can understand. Application generators have three major benefits:

1) They save time. You can write software faster.

2) They are perfect for quickly demonstrating an application.

3) They can often be used by non-programmers.

App Gens have two disadvantages:

1) The code they produce is often not as efficient as the code produced by an experienced programmer.

2) They are often limited in what they can produce.

Application generators tend to be general purpose tools. Alternatively, they may be very specific, providing support for specific applications, such as connecting voice response units to mainframe databases, voice messaging system development, audiotex system development, etc. Application generators are often used in programming voice processors.

One of the application generators' bigger advantages is their ability to translate user specified screens and menus into programming code. In essence, you produce the screen or menu using an interface as simple as a word processor. Then the applications generator translates that screen into programming code in a language, such as "C." Once translated into "C", a proficient programmer could go through the code and "improve" on it.

Using Graphical VOS To Create Fax Application

Parity Software's Graphical VOS combines flowcharting functionality with their classic VOS programming language. Graphical VOS supports the major Dialogic and Dialogic-compatible fax boards using simple function calls. Dialogic-compatible fax boards include those from Gammalink and Intel. Parity Software has been shipping VOS, an applications-oriented language which is similar to use as Visual Basic since 1989. Graphical VOS allows fax application developers to rapidly develop the application using the flowcharting interface. For the programming types, you can just program in VOS directly.

VOS simplifies fax development because the product supports every feature on every Dialogic or Gammalink fax card. Therefore you don't need to call to languages like C to add features to your application. VOS provides a "multi-tasking" environment where several VOS programs can be executing at the same time, allowing you to build hi-density applications without taxing your computer. VOS is cross-platform compatible, so you can write your application on, NT for example, yet deploy it on Unix, DOS, Windows 95 or NT.

To send a fax using a Gammalink board requires just a few function calls.

The following VOS program faxes the C:\AUTOEXEC.BAT file:

```
program
    sc_wait(1);
    sc_offhook(1);
    Gset(1, 2, "C:\AUTOEXEC.BAT");
    Gstart(1, 1);
endprogram
```

As another example, the following function call is all that is required to receive a fax and store it in a TIFF file using a Dialogic FAX/xx board:

```
fx_recv(line, "FAX.TIF", 1);
```

The three arguments to fx_recv are:

1. The FAX/xx channel number.
2. The file name to store the fax (may include drive and path components).
3. The file type. This may be 0 for a "raw" fax file, or 1 for a TIFF/F file.

To give an example of a complete VOS program which uses a FAX/xx board (could be a FAX/40 or FAX/120), the following program waits for a call from a fax machine, receives a fax, then re-transmits the fax to a given telephone number. This is a simple example of "fax store-and-forward":

```
dec         # Declare variables, name : length (char)
    var line : 2;
    var filename : 12;
    var code : 3;
    const FAX_PHONE = 5551212; enddec program
    line = 1;                        # Hard code line 1 test
    sc_trace(line, 1);       # Trace speech card events..
    fx_trace(1);             # ..fax events for debug      file-
```

```
name = "FAX" & line & ".TIF";
    sc_onhook(line);
    sc_wait(line);          # Wait for a call
    sc_offhook(line);               # Answer call
    sleep(10);
    fx_setstate(line, 0);  # Become a Called
    code = fx_recv(line, filename, 1);
    if (code <> 2)
    vid_write("Receive failed");
    restart;
    endif
    vid_write(fx_att(line, 6),"pages received");
    sc_onhook(line);                # Disconnect call
    sleep(30);
    sc_offhook(line);               # Start new call
    sleep(20);                      # Wait for dial tone
    sc_call(line, FAX_PHONE);
    if (sc_getcar(line) <> 10)
        vid_write("Call not completed");
        restart;
    endif
    fx_setstate(line, 1);  # Become a Caller
    fx_file(line, filename, 1, 0, -1, 0, 0, 0, 0);
    code = fx_send(line, 0);
    if (code <> 1)
        vid_write("Send failed");
        restart;
    endif
    vid_write(fx_att(line, 6)," pages sent");
    restart;
endprogram
```

The functions named sc_... are used to control a Dialogic speech card such as a D/41 or D/121 which is used to answer the call, and could in addition be used to play menus to the caller and collect touch-tone responses in order to select a document. Some typical sc_ functions include:

sc_onhook(line)
Puts line on-hook ready to receive a new call.

sc_wait(line)
Waits for a new in-coming phone call on the given line.
sc_offhook(line)
Answers call by going off-hook ("picking up the phone").
sc_play(line, filename)
Play a pre-recorded audio file to the caller.
sc_call(line, phone_nr)
Dials a phone number and performs call progress analysis to determine if the call was answered or if a busy tone or other result occurred.

The important fax-related functions are:

fx_setstate
Determines whether the fax board should behave as a Caller (the device which initiated the call), or a Called (the device which received the call).
fx_recv
Analogous to sc_record, saves received fax data to a given file name.
fx_file
Opens a given file in preparation for a fax transmission. Several files may be included in one transmission.
fx_send

Analogous to sc_play, transmits fax data from one or more files opened by fx_file.

The function sleep is used to suspend a VOS program for a given length of time, expressed in tenths of a second. The vid_write function writes to the PC screen, and is handy for displaying messages.

For another example, the following complete VOS program faxes the AUTOEXEC.BAT file using a CAS-compatible board such as the Intel SatisFAXtion:

```
program
    Fsub(1, 5551212, 0, 0,"",
"C:\AUTOEXEC.BAT");
endprogram
```

This chapter was provided by Bob Edgar, founder of Parity Software Development Corporation. For more information about VOS, buy Bob's book: *PC Telephony* from Flatiron Press, 800-LIBRARY. I thank him for his help and advice with this chapter.

For more information, visit: www.paritysw.com

Chapter 12

Computer Telephony Busses

Introduction

What are computer telephony buses? A bus is a familiar term in the computer world. It means a electrical, mechanical, and signalling protocol to allow different components within a computer to work together. Examples of buses in the computer world include ISA bus, EISA bus, CompactPCI, sbus, VMEbus, etc.

The computer telephony industry has adopted the use of buses for the same reason—to allow different computer telephony components to work together in the same system. Unlike the "host bus" that allows these expansion boards to communicate with the computer's CPU, a separate bus for carrying real time information between various computer telephony devices is critical to making these devices perform as a complete system. In a standard IBM

PC environment, these buses are physically implemented as a mezzanine bus, and the physical connect consists of one or multi-drop or point-to-point ribbon cables connecting the computers together. On other platforms, the physical transport could be the same at the host bus, like the VME platform or a proprietary switching platform.

This chapter will explore the four different kinds of buses that have arisen in the computer telephony world—the Analog Expansion Bus (AEB) (tm), the PCM Expansion Bus (PEB) (tm), the Multiple Vendors Interpretation of Protocol (MVIP) (tm) bus, and the Signal Computing Bus (SCbus) (tm).

Common Elements

Despite the different architectures and protocols of the four standard buses, all were designed to accomplish the same purpose. Each computer telephony bus allows the system integrator to extends the capability of a single computer telephony component by allowing multiple components to work together in the same telephone call. There are two basic reasons for this:

The first reason to use a ct bus is when the telephone network interface is physically distinct from the signal computing component (voice processing, fax, speech recognition, etc.). There are many ways for a computer telephony system to connect to the public switched telephone network, including analog loop start or direct-inward-dial, and digital T-1, E-1, basic rate ISDN, primary rate ISDN, or proprietary PBX links. By separating the network interface from the signal computing component, system integrators can mix and match different components to build the system they need. In addition, different vendors can supply different components of the total system.

The second reason to use a ct bus is when the system integrator wants to augment the capabilities of computer telephony system. For example, some installations might call for speech recognition capabilities; others may not. By employing a ct bus, both types of systems can use the same basic architecture and most of the same components. That system that requires speech recognition simply augments its system configuration by adding a speech recognition resource via the computer telephony bus.

In addition, the new generations of buses, like the SCbus, have a large inherent switching capability which greatly increases the switching fabric within a computer telephony system, and in many cases can be used to replace an external switching device.

Analog Expansion Bus

History

The AEB is the first open computer telephony bus, and was introduced by Dialogic Corporation on their D/xx line of voice processing products. In order to spur industry growth, Dialogic publishes the specifications to the AEB, and provides technical document for the bus at a nominal charge. The AEB was designed to allow system integrators to attach Expansion Modules, like a facsimile component, that extend the basic capabilities of the D/xx, and to allow system integrators to attach digital Network Interface Modules for use with the D/xx.

The AEB was eventually adapted by the other leading voice processing manufacturers—Rhetorex and Natural Micro-Systems and by fax vendors GammaLink and Brooktrout.

Functionality: The AEB supports four voice processing channels via a 20 pin connector. Each of the four channels has a audio signal, and signaling transmit and receive signals.

The audio signals are terminated, 2-wire, bi-directional access points. The signaling states are carried on TTL logic level signals to allow software-transparent (to the application program) signalling connections between different computer telephony modules connected to the AEB. On/off hook commands and network alerting (ringing) or off-hook complete (using T-1 E&M protocol) signals are passed via the signaling pins.

Advantages and Disadvantages

The AEB is simple, inexpensive computer telephony bus that has been adopted by the leading computer telephony for low density systems.

It is not well suited, however, for high density systems. The analog nature of the bus does not lend itself to the multi-drop or high density nature of many larger computer telephony systems, which is why Dialogic introduced the PCM Expansion Bus.

The two most common configuration examples are: a) Voice processing card to resource module (fax or speech recognition) and b) Digital network interface to resource modules.

PCM Expansion Bus (PEB)

History

The PEB was introduced by Dialogic Corporation in 1989. It is the first digital, open computer telephony bus and it helped the industry expand into higher density systems. The PEB used time division multiplex techniques to pass digital, pulse code modulated date (hence its name) between computer telephony computers.

In order to spur industry growth, since 1991 Dialogic has published the interface specifications to the PEB and does not require a license for its use. Detailed technical documentation to the PEB is available from Dialogic at a nominal fee. Given Dialogic's market position, a large number of computer telephony component providers have PEB based components.

Functionality

The PEB is a 24 pin, high speed, digital, TDM bus that, depending on the clocking rate, can support 24 or 32 simultaneous telephone channels. To support T-1 and 24B+D PRI applications, the PEB is clocked at 1.544 Mbps and supports 24 full duplex channels. To support E-1 and 30B+D PRI applications, the PEB is clocked at 2.048 Mbps and supports 32 full duplex channels.

The PEB world is divided into two types of components—Network Interfaces attach to the telephone network, and typically control the clocking of the PEB, and Resource Modules that perform some signal computing function (e.g. voice processing). Resource Modules switch onto and off of the PEB via time slot assignment—each resource module is capable of transmitting on and/or receiving from some range of time slots on the PEB.

Advantages and Disadvantages

The PEB is a widely embraced bus. Many companies have PEB-based product on the market, allowing developers to mix and match pieces to serve a broad range of applications. The PEB is relatively easy to implement, and takes little material cost. Dialogic offers the technical specifications for the PEB at virtually no cost, and offers design assistance.

The PEB, though, is a first generation bus. It relative lack of time slots is a disadvantage in high density systems.

Configuration Examples

The most common configuration example is that of a network interface connected to one more resource modules. Since each resource module can select transmit and receive time slots, different resource modules can participate in the same call, even at the same time. For example, a T-1 network interface (e.g. DTI/211) might be connected to 24 channels of voice processing (e.g. D/240-SC), four channels of fax (e.g. CP4-SC), and 12 channels of speech recognition (e.g. VR/120p). Depending on the needs of the caller of the system, the appropriate resource can dynamically connect and disconnect on the call.

Another popular configuration is the drop-and-insert configuration. In this configuration, the call processing unit fronts end a switch. The front-end call processing unit intercepts some or all of the calls destined for the switch, and provides call processing resources for the call. The call processing unit might, for example, query the caller for an account number or offer to automatically fax information to the caller. If the caller needs to reach a live operator, the call processing unit can pass the caller through to the back end switch, and on to an operator. Any information the call processing unit collected from the caller, like an account number, can be passed to the operator or an associated computing device.

To implement a drop-and-insert configuration, a network interface is placed on both sides of the PEB cable, both logically and physically. In between, there is a cross-over cable. Automated call processing equipment can be placed on one or both sides of the cross-over cable.

MVIP

History

The Multi-Vendor integration Protocol was announced in 1990 by Natural MicroSystems along with Mitel Semiconductor, GammaLink, Brooktrout Technologies, Voice Processing Corporation, Scott Instruments and Promptus Communications. MVIP was presented as an open standard for interoperability among telephone-based resources, including trunk interfaces, voice, video, fax, text-to-speech, and speech recognition, to enable design of larger, multi-technology computer telephony systems.

Functionality

MVIP defines intra-node and inter-node digital communication buses and defines multiple switching models for allowing devices to communicate in real time.

Intra-Node Bus

The MVIP Bus specification is based on Mitel's ST-Bus, and is a digital telephony bus that carries PCM data between devices. The MVIP Bus consists of 16 serial data lines clocked at 2.048mhz, providing 512 half-duplex timeslots, or 256 full duplex timeslots. Since transmit and receive timeslots are allocated in pairs to each resource (channel on a device), the MVIP bus can support nonblocked communication among up to 256 resources.

The bus can be implemented on ISA, EISA, or MCA platforms as well as other computing platforms. In a PC implementation, the MVIP bus would be implemented using a ribbon cable with a 40 pin connector. If it is necessary to transfer signalling information between devices, the MVIP recommendation is to carry this information through associative channel signalling protocol, using data timeslots. In this case, the number of available timeslots for data transfer is further reduced to 128.

Multi-Node Systems

MVIP defines a specification for interconnecting individual nodes into larger, distributed systems. Called Multi-Chassis MVIP, this specification was presented in 1993 and currently proposes 4 different implementations. These are as follows:MC-1: up to 1408 timeslots running at 4.096 Mbps over twisted-pair cableMC-2: up to 1536 timeslots over FDDI-II or copperMC-3: up to 2300/4600 timeslots over SDH/SONET fiberMC-4: (ATM).

Switching

The MVIP architecture assumes that most traffic in most MVIP applications is being connected between network interfaces and call processing resources. Devices are differentiated by whether they have telephony interfaces or whether they are doing processing only (ie, voice store and forward, fax, or ASR). In the 3 switching models originally specified in MVIP, a telephony interface device or a switching device was required to have some level of switching capability and that this device would switch information between the various resources involved in processing a call.

The level of switching affects the maximum system size and configuration. The three levels are:

MVIP Switch Compatible

The lowest level of switching requires that the device maintain a switch capable of making full-duplex connections of any incoming telephony channel to a subset of MVIP bus timeslots and vice versa. In this case, a device would only be capable of switching information among a limited set of resources.

MVIP Standard Switching Compliant

This level of switching allows the device to switch data from the network interface to any MVIP bus input or output timeslot consuming only 4 timeslots on the MVIP bus (2 transmit and receive pairs). In order to switch information to another channel on the same device, an MVIP Standard Switch Compliant device would use 2 MVIP bus timeslots.

MVIP Enhanced Compliant Switching

An Enhanced Compliant Switching device is capable of switching data from the network interface to any MVIP input or output timeslot. Because it has additional switch paths to connect incoming and outgoing network channels within a trunk interface, an Enhanced Compliant device can connect to any other channel on the same device without consuming any MVIP bus timeslots. An enhanced switch compliant device could serve as a central switch for an entire MVIP chassis.

With the new FMIC (Flexible MVIP Integrated Circuit) chip, it is possible to have switching completely distributed so that information could be switched directly between resources. However, this is not recommended in the specification and at present many devices do not yet use the FMIC.

Advantages and Disadvantages

Advantages: Higher capacity: The MVIP Bus offers higher capacity than the PEB which requires a separate switching device for more than 24 or 32 channels.

Choice of Implementations

Telephony interface manufacturers can support different levels of switching compliance depending on system size and configuration. This allows them to economize on cost depending on the intended application.z Products: MVIP claims approximately 100 hardware products are available.

Disadvantages

Timeslot allocation: Because MVIP requires fixed transmit and receive timeslots for each processing resource, certain applications like broadcasting data to multiple resources also requires multiple timeslots to duplicate the data for each listener. Depending on the level of switching compliance, additional timeslots are required to route calls between telephony interfaces for configurations like drop and insert.

In-band signalling: Signalling information or messages between MVIP devices must be carried on the host bus or in-band using data timeslots. If this information is carried in band, this halves the effective system capacity.z Lack of standard APIs: Because MVIP does not define a standard software architecture, it is difficult to design complex systems using products from multiple vendors, each of which use proprietary APIs.

Sample products that support the MVIP (Note: this is not a complete list. For a comprehensive list of actual hardware products and the level of switching compliance, contact Natural MicroSystems.)

For the latest specifications and information, visit: www.mvip.org

THE SCSA Hardware Model

History

The Signal Computing System Architecture (SCSA) is a hardware and software architecture launched in March 1993 by Dialogic Corporation and 70 other telecommunications and computing equipment suppliers including Digital Equipment Corp., Northern Telecom, NEC, and IBM to simplify the task of building larger, more complex computer telephony systems using multiple technologies. SCSA active supporters currently number 240+ companies, many of whom are participating in defining the SCSA specifications.

The SCSA Hardware Model defines the hardware portion of the SCSA architecture. It consists of a real-time digital communications bus with a separate messaging channel, a multi-node network architecture for development of multi-node systems, and a single, distributed switching model. Servers using the SCSA Hardware Model can be controlled using various programming models including the SCSA Telephony Application Objects (TAO) Framework or through other vendor-specific software models.

Intranode Bus

The SCbus is TDM bus for computer telephony consisting of 16 synchronous serial data lines for real time communication among devices in a single node and a dedicated messaging channel for carrying signalling and messages between devices. The SCbus supports 1024 bidirectional 64 Kbps timeslots in a mezzanine bus implementation running on a ribbon cable, or 2048 time slots when running in a backplane implementation. An SCSA hardware implementation specification for the VMEbus has been endorsed by VITA (the VMEbus industry trade association) and is currently in submission to the American National Standards Institute (ANSI). The SCbus can also be adapted to other platforms, for example, as part of the backplane of a proprietary switch.

The optional messaging channel provides a means for realtime communication of messages and signalling information among resources "out of band." The message bus allows devices to communicate this information directly without going through the host processor or application for faster system response time. Additionally, developers can integrate specialized resources that use the message bus to get command control and status information. Communication over the message bus is faster and more efficient than embedding the message handling in the data stream and does not consume data timeslots of the SCbus.

The message channel can be implemented using any transport capable of transferring data in real time. For example, a developer could implement the message channel over an Asynchronous Transfer Mode (ATM) network where certain packets would be dedicated to messaging.

Multinode Expansion

The Multinode Network Architecture (MNA) provides the ability to connect multiple nodes to build large systems. The MNA is application independent; its connections are transparent to the application. The MNA provides the capability to incorporate nonSCSA nodes. For example, it can be used in high-density PBX integrations to provide a gateway function that extends the SCbus to other nonSCSA products. The MNA can be implemented through various media, depending on the system size requirements and physical location of nodes. Below are 2 examples:

SCxbus

Connects up to 16 co-located nodes using a ribbon cable, for systems up to 1344 ports. This implementation is physically compatible with MC-1.z ATM: An ATM switch can be used to connect larger numbers of geographically distributed nodes for multiple thousand line systems. In this case, the messaging channel can be implemented through dedicating ATM packets for this information.

SCSA Switching Model

SCSA defines a single, completely distributed switching model in which all devices are peers. It features flexible time slot allocation where all devices can transmit or receive on any time slot on the bus, enabling communication between any devices in the system. Other features include automatic clock fallback and switchover for fault tolerant application requirements, time slot bundling with full frame buffering for technologies like video, and broadcast capability where one device can transmit to multiple devices while consuming only 2 timeslots regardless of the number of "listeners." This model is easily scalable and suitable to implementation on different hardware platforms, for example, from a PC ISA bus to VMEbus.

Switching is controlled through the SC2000 chip, designed by Dialogic and built by VLSI Technology. The SC2000 also offers compatibility modes for other buses including PEB, MVIP, and ST-BUS, allowing developers to design hardware that with different software loads can operate in SCSA, PEB, or MVIP system.

Advantages and Disadvantages

Advantages: Higher capacity: The SCSA Hardware Model today offers the highest capacity and fastest speed of any computer telephony architecture, for the greatest degree of scalability.

Compatibility modes: Because the SC2000 chip offers compatibility modes, developers building hardware devices for multiple hardware platforms can economize by building a single hardware product that can run in different modes, with the appropriate software.

Single switching model: Because all SCSA products implement the same switching model, there are no limits on device location, enhancing system scalability. It also provides the most efficient allocation of timeslots and eliminates the need for dedicated switching hardware.

Message channel: The separate message channel provides faster system response time especially in client server configurations where applications may be running on a remote platform. The message channel also supports applications where out-of-band signalling capability is required without consuming any data time slots.

Defined software architecture: Because SCSA also defines a comprehensive software architecture (SCSA TAO Framework), the hardware architecture is designed with consideration for application development concerns like performance features required for client server environments. SCSA TAO Framework offers several capabilities critical to supporting distributed computer telephony systems like the ability to support multiple applications, dynamic resource allocation, and standardized vendor-independent application programming interfaces.

Disadvantages: As a newer architecture, components that support the SCSA Hardware model are still being released. Several companies are still designing products which will be released in the upcoming months.

For the latest specifications and information, visit: www.scsa.org

Summary and Recommendations

Computer fax board buyers should give careful consideration to the architectural platform when choosing a product supplier. Although many computer telephony technologies are still in an evolving state, the demand for integrating these "media processing services" with facsimile processing is growing, because there is a fundamental need to make "bridges" between once separate communication tools like telephones, facsimile and workstations in order to bring information to users in the most useful form.

In choosing an architecture, the user should also evaluate the software architecture, which can make a great difference in the time needed to bring their systems to market. The software architecture must be able to support client server considerations like support for multiple applications. Physical connectivity between hardware alone is of limited value if the products cannot provide features like multi-application support and dynamic allocation and sharing of resources.

Chapter 13

Fax Modems

Introduction

There is much confusion regarding what the perplexing varieties of Class 1 and Class 2 data/fax modems can and cannot do. There are claims and counterclaims as to whether they are capable of working in a busy fax server environment, where mission-critical faxing takes place.

Anyone considering the use of Class 1 or Class 2 modems for mission-critical faxing must consider the pros and cons of these technologies before making a final decision that may affect an important aspect of the running of a corporations business.

Class 1 Modems

Class 1, which was approved by the ITU-T in 1990, was one of the first specifications for fax communication. It is a series of basic Hayes AT commands used by software to control the board. The first layer, which is addressed by Class 1 modems, operates close to the data link level. At this level, only very simple operations are performed: HDLCs (High-Level Data-Link Control standard.)

As a fax handshake signal format, HDLC frames are used for the binary coded handshaking. They may have a digital identification signal (DIS) describing the machines features, a non-standard facilities (NSF) frame describing vendor-specific features, and a called subscriber identification (CSI) frame containing the answerers telephone number.

A Class 1 and 2 modems contact with the world is through an RS-232 connection serial port operating at a rate of 19.2 kbps or less. This means that every time a byte of data is sent, it must go to the serial port and there are delays inherent with that sending 8 bits at 19.2kbps takes time. It takes even more time to send an HDLC frame or to do the T.30 protocol. For example, when the fax machine at the other ends sends it's DIS, it is received by the modem, which then transmits it over the serial port to the computer, the computer then interrupts whatever it is doing, interprets it, then sends the appropriate command through the port back to the modem. It is important to understand here that the T.30 protocol does not reside in the modem, but in the computer, and that its processing requires the computers undivided attention if it is to direct the modem at each step of the fax communication process.

The Standard That Never Was

When, in 1991, Class 2 was published as a ballot standard by the ITU-T committee, the ballot failed due to a number of technical and political issues. There was much disagreement over what members and industry believed the specification needed to make it work right. The committees debated many issues for over a year before agreeing on what would be in the final Class 2 specification.

Unwilling to wait for the ponderous ITU-T decision-making process to grind on, a number of manufacturers anxious to get into production, took the ballot as if it were a specification and went into production. The result was that after a time there were sufficient Class 2 modems out in the world to make the ballot a de facto standard, although several key areas are undefined and can vary from vendor to vendor.

After the ITU-T solved the issues, they published the final Class 2 specifications, which differed significantly from the Class 2 ballot. In order to distinguish the new real Class 2 specifications from the earlier ballot ones, the ITU-T used Class 2.0 as the new name. Thus there is a fundamental difference between Class 2 modems based on the ballot standard and Class 2.0 modems based on the official ITU-T standard.

One of the most important differences deals with the packet layer protocol in the Class 2.0 specification, guaranteeing more secure throughput over the RS-232 line. Class 2 does not feature this.

Programmers began creating software for the Class 2 modems precisely because they did (and do) exist in great numbers. However, when they want their software packages to be retroactively compatible, programmers discover there is no source document that they can work with, except perhaps for a photocopy of what used to be the interim ballot and nothing else. This is because Class 2 was never approved by the ITU-T. In fact, it is not even possible to get back copies of the original ballot, because once the committee approved the final Class 2.0 specification, all previous work was discarded.

So Class 2 is an illegitimate offspring nobody wishes to recognize and with the Class 2 installed base still out on the world, these questions and problems will not soon disappear.

In the committees view, then, Class 2.0 is Class 2 done right, and it has the ITU-Ts blessing and official industry approval as a standard; that is, there is an official specification available. Unfortunately, since it took the ITU-T over a year to come up with this standard, by the time it was published it was already obsolete in terms of new features added to T.30, which ought to have been included. Not all of the latest fields such as subaddressing and binary file transfer protocols are present in the official Class 2.0 specification.

On the positive side of this equation, the next revision, Class 2.01 or 2.1, requires only fairly straightforward changes, such as putting in commands to support the latest features of T.30. This activity is more in the line of mechanical editing than design, and should not excite controversy. It comes down to deciding which command will support a new T.30 option and selecting the right letter for it.

The problem, however, remains that the installed base of Class 2.0 modems is still relatively small, especially when compared to the enormous installed base of Class 2 devices that are already out there.

Class 2 Modems, T.30, and Compatibility

A Class 2 (and 2.0) modem has the T.30 protocol on board. It will do such things as SET DIS and SET CSI. The user can command it to connect to the other station, and it will dial the telephone number, exchange the DIS, the DCS, and preform all of the handshaking operations before sending the data. There is no need on the hosts part to send HDLCs or image, just data. The Class 2 modem will perform the fill bit operation and other T.30 functions that may be needed.

This has a plus and a minus. The plus is that the modem does the T.30 handshaking, with time-sensitive and time-critical functions being performed by it instead of the computer. The minus is that the T.30 protocol is locked in the hardware, therefore any changes to the fax protocol cannot be done simply through software, requiring instead new hardware. In a server environment, image-processing such as

ASCII-to-fax, headers, PCX, etc., must be handled by the host.

Another problem is that many Class 2 fax modems are incompatible with one another and other fax devices because there is little cooperation among fax modem manufacturers. This is a key point and must be given careful consideration.

The T.30 protocol has some areas the duration of timeouts, for example that are not specifically defined and are therefore subject to interpretation.

During the 1984 to 1986 time frame, there were very few T.30 fax protocol engine engineers in Japan, and none in the U.S. Although they competed on manufacturing, performance, and price, all Japanese fax manufacturers cooperated with each other to ensure that all manufacturers could send and receive faxes to and from other machines. Precisely because it is open to interpretation, this rendition of the T.30 protocol was not tampered with, ensuring compatibility. Since there was only a handful of manufacturers, this was not too difficult an enterprise. Now, however, it is an entirely different story.

At present, the majority of those involved in the manufacture of fax modems have produced fax protocol versions based on their individual interpretations of the T.30 recommendation. Although this may occasionally result in what is boasted of as a superior fax protocol, the problem still remains that it will not talk to all the fax devices machines and boards out in the world.

There is a significant number of incompatibilities for Class 1 and Class 2 modems and the general population of fax devices in the marketplace. End-users and even developers have very little recourse, unless they actually write T.30 code themselves, but then they are burdened with that task of fixing (and what may need fixing is not always obvious) something that does not work at all or only partially. In the end it may still be incompatible, only now with an entirely different set of fax devices. The level of T.30 compatibility testing that the fax board and machine companies carry out is unknown in the Class 2 fax modem environment.

The problem has been exacerbated by increasing competition. Clearly, from a compatibility point of view, two fax modems from the same manufacturer can probably talk to each other, but there still may be other problems such as host computer dependencies and timing issues. (For example, a modem may not operate exactly the same in different computers.) Timing issues are critical because, as asynchronous device, a modem expects data when it requests it. If the host computer is not immediately there to provide it and a piece of data is missed, a scanline or error correction mode (ECM) train dropped, the damage is done and the fax does not come out straight or, if ECM is present, it becomes necessary to retransmit.

ECM is available in Class 2.0, but not in Class 2. This is one of the reasons why many developers working with Class 1/Class 2 modems generally drive them in Class 1 format for control (Class 2 supports Class 1 operations), and do all T.30 operations on the host computer.

With Class 1 almost everything is done by the host computer, where as with Class 2 some of it is off loaded to the modem, such as some of the buffering and timing responsibilities.

Multiple-Modem Applications

A Class 2 modem has the capability to buffer a few seconds of data. So if some data is missed, it just slows down a bit while making it back up. Because of this, a Class 2 is better able to handle a busy host, and it would be theoretically possible to run multiple Class 2 modems on a single host and a fax server. It is extremely unlikely to run multiple Class 1 modems because of the interrupts the need to service the data is almost constant.

With a Class 2 modem the problem of needing to adjust to the capabilities of the remote device is not solved. The information about various compressions (MH, MR, MMR) is still on the host and carried out by it, and it has to provide the right data to the modem. So steps such as conversions from ASCII into fax are still being performed by the host, through a serial port.

Serial Ports

A DOS serial port is interrupt-driven (and therefore slow for fax purposes), with each interrupt occurring on a character-by-character basis. To properly keep up with fax it is necessary to run at a greater rate: one equal to the faxes. If the rate equals the fax speed, eventually this results in underrun situations. Another point to consider is that, as an asynchronous interface, a DOS serial port uses ten bits per character instead of eight. These additional two bits, a start and stop bit, can consume considerable CPU time. A fax transmission running at 14400 bits per second requires just that: 14400 bits every second. However, if a DOS serial port is used, that means 14400 plus a 25 percent timing overhead to accommodate start and stop bits.

This means the CPU is doing the interrupt of service routine very often, (every 100 msec or so) to read in the next character to the serial port (for each modem).

Therefore, any CPU activities that are time-critical, such as switching memory banks, going in and out of 486 or Pentium modes, etc., which Windows is very prone to do, can severely impact the faxing as well as the program processing capabilities of the modem and computer. Also, any other operations that are interrupt-free, such as screen rights or display updating, may cause the serial port to experience severe character losses. The interrupt overload also degrades a PC's general performance by forcing it away from other tasks to service an interrupt and then returning to continue the previous task. This kind of back and forth and back and forth switching carries with it considerable overhead.

Conclusions

What an end-user needs, must be determined by use; how many faxes a day, a week, a month are sent and received. For the desktop, laptop, or casual user there is nothing wrong with Class 1 or Class 2 modems. There may be compatibility issues but there is usually a fax machine handy if a fax has a problem in getting through. Most of the Class 1 and Class 2 fax modems sold today have never sent or received a fax. They were bought for data. In fact, nowadays, anyone looking for a data modem is hard-pressed to find one that does not offer a fax capability of some sort. Computer store salesmen are fond of saying, "buy a modem and get fax for free." This, however, is one time when it may be useful to look that gift horse in the mouth.

System integrators and ISVs build mission-critical fax applications. If some of their customers send hundreds or thousands of faxes a day, it is crucial that the faxes get through, and that the user know if a specific fax did not, and if not, why. These users cannot afford to deal with incompatibility tie-ups. For vital applications of this kind, the price of the fax board, whether $139 or $1,000 becomes a much less significant variable in the systems cost. That is the market high end fax board address: one in which getting the fax through is critical, where keeping track of expenses is critical, and where sending faxes at the lowest transmission costs as in the case of high-volume applications is an important goal. In terms of the occasional fax sent from the laptop or standalone PC, the user maybe better off by buying a data modem and getting fax for free, always remembering, however, that one gets what one pays for!

Section V

Fax Hardware Intelligent Fax Boards

Chapter 14

The Brooktrout Fax Hardware Approach
and Product Line

Brooktrout is a leading vendor of multi-channel fax boards. Introduced in 1992, and enhanced continuously since then, Brooktrout's TR114 Series Universal Port fax and voice boards are designed for high performance fax and voice messaging systems. They are widely used for fax-only and mixed fax-and-voice applications by service providers and enterprises for applications such as fax broadcast, fax-on-demand and LANFax servers.

Brooktrout's C-language API—the Bfv (Brooktrout fax and voice) API)—offers developers a well integrated universal port API encompassing call control, fax processing and voice processing. The wide range of configuration in which the TR114 Series is available, including many different telephone network interfaces, PCI and ISA bus models and boards ranging from two to sixteen channels per port, make it easy and cost-effective to deploy systems worldwide in the appropriate configurations.

The TR114 Series supports virtually every advanced fax feature, enabling the fastest transmission and the greatest file compression. This experience translates into superior rates of connectivity with fax devices worldwide. Taken together, these capabilities offer lower operating costs to network operators.

In 1997, Brooktrout announced its next-generation BOSTON architecture for development of its future universal port products. BOSTON offers developers a will integrated development environment for fax, voice and data products. Brooktrout also pledged to protect application developers software code by offering a forward migration path into the BOSTON architecture for existing applications, a capability that was delivered in December with the release of the Bfv API version 4.0. Brooktrout also aims to use BOSTON to provide a range of cost effective hardware platforms by developing board-level firmware that is easily retargeted to different DSP and microprocessor environments. This will enable Brooktrout to use a variety of DSPs and microprocessors to deliver a range of cost effective, high performance boards.

TR114 Series

The TR114 Series Universal Port boards are designed for high performance fax and voice messaging systems, such as those used by telecommunication service providers, messaging system vendors and network communications server vendors. The TR114 Series boards offer full fax and voice processing on each channel of a single multichannel board. This unique capability increases system reliability and provides unequaled flexibility to system developers.

At the heart of each Universal Port board are software controlled digital signal processors. Each port of a TR114 Series board has a dedicated DSP with and embedded 32-bit control microprocessor. This processing power ensures that each channel's function is totally independent and that all resources are available on every channel. The TR114's functions are provided by downloadable firmware. Additional functions and features can be added later with a software upgrade.

The TR114 Series Universal Port board supports all leading fax and voice features.

Fax

- 14.4 kbps transmission and reception, with automatic step-down
- MMR, MR and MH compression
- Error correction mode
- On-the-fly file conversions of ASCII, TIFF, PCX/DCX to G3 fax
- On-the-fly combination of files
- On-the-fly page size conversion and scaling among A4, A3 and B4 page sizes
- On-the-fly forms overlay (up to 4 different forms per channel)
- Multi-line headers with selectable fonts

Call processing

- Support for DTMF and pulse dialing
- Advanced international call progress tone detection and analysis
- Precise, accurate line status reporting

Voice

- 24 and 32 kpbs ADPCM
- 64 kbps mu-law
- High performance cut-through and talk-off
- Static and dynamic adaptive echo cancellation

The TR114 Series is distinguished by the wide range of worldwide telephone network interfaces it supports. Boards are available with on-board loop start, direct-inward-dialing (DID), basic rate ISDN (BRI) and, most recently T1 robbed bit signaling interfaces. In addition, Brooktrout offers stand-alone network interface boards for T1 robbed bit signaling and primary rate ISDN telephone networks.

Brooktrout has also been a leader in the range of configurations it offers. Recently, it broke new ground in high density products by reducing the number of PC slots per T1-span to two with the introduction of the TR114+I8V-T1 (an eight channel board with an integrated T1 network interface board) and the TR114+I16V (a sixteen channel board). The full line of TR114 Series products is summarized in Table 1.

TruFax

Brooktrout's TruFax Series fax boards are computer-based fax boards designed for small to medium scale general purpose fax servers and systems. TruFax is an entry level product offering two fax channel per board and up to 12 channels per server. The TruFax 200 incorporates the same advanced call progress and T.30 functions used in the TR114 Series to ensure reliable fax transmission worldwide. TruFax, however, supports fewer advanced fax features than the TR114 Series.

Brooktrout's TR114 Series Product Line Overview

Bus Type & Network Interface or TDM Bus	Channels/board	Products Approved for Use In:
ISA Loop Start	2: TR114+I2L 4: TR114+I4L	US, Canada, Hong Kong, Japan, UK, Australia, Norway, Netherlands, Singapore, Germany, Sweden, France, Switzerland, New Zealand, Denmark, Ireland, Italy, Spain, Czech Republic, Malaysia, Thailand, Argentina
PCI Loop Start	2: TR114+P2L 4: TR114+P4L	US, Canada
ISA Basic Rate ISDN	2: TR114+I2B 4: TR114+I4b	UK, Netherlands, Norway, Germany, Sweden,France, Switzerland, Denmark, Austria, Belgium, Finland, Greece, Ireland, Italy, Portugal, Spain, Iceland
ISA DID or Combo (DID & Loop)	2: TR114+I2C/D 4: TR114+I4C/D	US, Canada, Hong Kong
ISA MVIP or SCbus	8: TR114+I8V/S 12: TR114+I12V 16: TR114+I16V	Worldwide
Network Interfaces boards: T1 Robbed Bit Signalling	TRNIC 8 channel TR114 with integrated T1 network interface: TR114+I8V-T1	US, Canada
Primary Rate ISDN Dual T1 ISA & PCI	PRI-ISALC-2T, PRI-PCI48V	US, Canada, Japan
Primary Rate ISDN Dual E1 ISA and PCI	PRI-ISALC-2E, PRI-PCI64V	UK, Australia, Netherlands, Germany, Sweden, Switzerland, France, New Zealand, Austria, Belgium, Denmark, Finland, Greece, Ireland, Italy, Luxembourg, Portugal, Spain

BOSTON

In July, 1997, Brooktrout introduced its BOSTON architecture to address many of the unmet needs of developers of fax and other electronic messaging systems. The electronic messaging market today provides many opportunities for developers of messaging systems and applications to deliver fax, voice and data messaging products and services to end-users at work and at home. As the demands of end-users increase and technology advances offer new capabilities, developers of electronic messaging equipment and messaging service providers push the envelope of electronic messaging platforms and development tools and place increasing demands on electronic messaging platforms and development tools suppliers like Brooktrout.

The Brooktrout Open System Telephony (BOSTON) architecture provides the framework for Brooktrout to deliver electronic messaging products that enable developers to deliver universal port systems more quickly and at lower cost. With BOSTON, Brooktrout will deliver:

- Graphical, object oriented, development tools for rapid application development
- Highly integrated C callable API libraries for universal port access to multiple messaging media types and call control abstracted from the telephony or data network
- Cost-effective and scaleable board-level platforms with integrated telephony network interfaces
- Extensible firmware to deliver electronic messaging capabilities that meet specific market and customer needs

BOSTON
Brooktrout's Architecture for Electronic Messaging

In seeking ways to help developers bring new systems and services to market quickly, Brooktrout identified that a layered, software-centric approach would provide significant advantages both to developers and Brooktrout. This resulted in a new vision for electronic messaging system development.

Brooktrout's Vision for Electronic Messaging System Development

Network Voice Fax

Layer	Attribute
Complete high-level development tool	Well integrated tools
Single API	Universal API with Same Look and Feel
Single Driver	Efficient/Portable
Firmware A / Firmware B / Firmware C	Extensible
Hardware	Flexible

The Brooktrout Open System Telephony architecture (BOSTON) was created to enable Brooktrout to deliver products that fulfill this vision. BOSTON is a software architecture that forms the framework for delivery of development tools and messaging platform components for electronic messaging applications. It is a full-featured, universal port architecture for multiple messaging media types which offers a high degree of scalability, extensibility, portability and configurability through the development tools and platforms which have been created from the architecture.

The power of BOSTON is based on its adherence to four core design principles:

- **Common interface for packet-based communication**

  Communication between firmware and various software

layers on the host, including the driver and API, is performed using data packets similar to packet switched networks. Packets abstract low-level commands and data types as "payloads" and offer the highest level of data communication flexibility. The core driver developed for BOSTON-class products will perform packet-switching and will provide seamless support of multiple application processes including management processes such as SNMP agents. In contrast to the traditional, hardware-centric, threaded driver, a packet switched paradigm for driver communications provides greater flexibility. In particular, it enables multiple processes, such as an application and a management agent, to address the same port simultaneously.

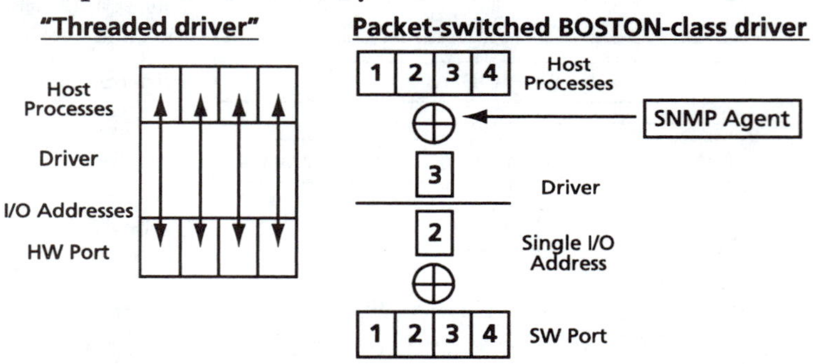

- **Structured parameter passing**

  Parameters are passed within a data structure between the application and the called functions as opposed to the traditional method of using an argument list. This allows function parameters to be abstracted.

- **Developed in portable C language**

  All software is developed in the C language which provides the highest level of portability between processors.

- **Highly modular software components**

  Firmware which provides the media processing and network and call control functions on the messaging platform is comprised of highly modular firmware facilities. Facilities are software modules that govern a very specific task such as G.721 voice coding, V.34 full duplex modulation, T.30 protocol processing or image resolution scaling. API libraries and drivers are comprised of modular components which allow extensibility of functions, independence from operating systems and abstraction from hardware.

The modularity of software is extended to the use of hardware and physical layer abstraction. Modules at the lower levels of software provide hardware and physical layer abstraction to make specific hardware components and physical layer interfaces transparent to most of the software and firmware. The following illustration depicts the abstraction of physical layer telephony interfaces.

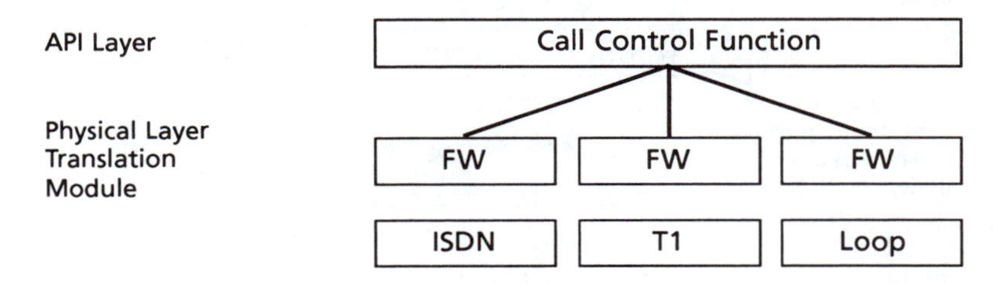

The BOSTON architecture delivers software components which take the form of development tools, application programming interfaces and drivers on the host PC and firmware which performs modulation, media processing, and call control on an add-in board messaging platform utilizing both DSP and scalar CISC processors.

New Products Based on BOSTON

Starting in the second half of 1997, Brooktrout began introducing new electronic messaging platforms based on its BOSTON architecture.

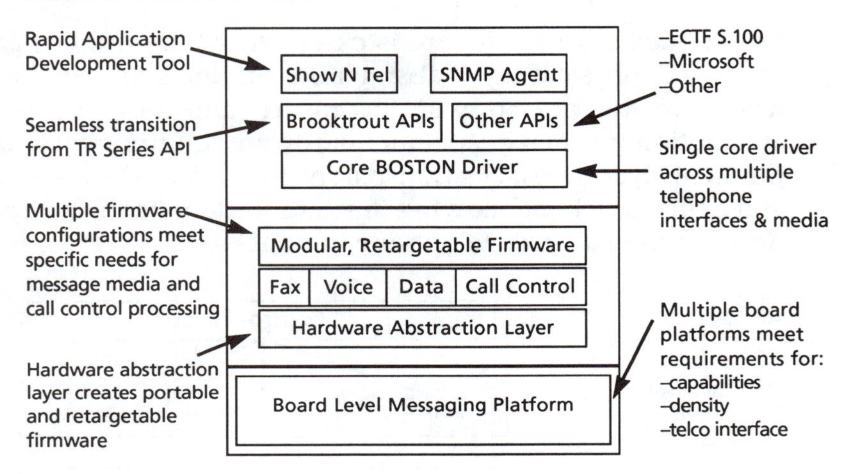

These platforms will provide fax, voice and data messaging capabilities targeted at the following applications:

- IP Telephony Platforms
- Data/fax Platforms for Network Servers
- High Density Fax Platforms for Enhanced Fax Services
- Scalable Voice and Fax Platforms for Unified Messaging

In the second half of 1997, Brooktrout announced new products for IP/Telephony (its TR2000), new ISDN network interface products, the first version of its BOSTON-style Bfv API, API version 4.0. It is expected that Brooktrout will announce BOSTON fax products in the first quarter, 1998.

For more information, visit: www.brooktrout.com

Chapter 15

The Commetrex Hardware Approach and Product Line

The Intelligent Fax Board Is Dead

The intelligent fax board is dead! Well, not really dead, but it's certainly no longer the fax board as you knew it. Other intelligent fax boards require fixed-function fax chips provided by DSP companies like Rockwell and National Semiconductor for each fax port. They do the job, but they're expensive and inflexible. This traditional approach to building fax or fax-voice systems out of these fixed-function boards is shown below.

<table>
<tr><td colspan="3">Application</td></tr>
<tr><td>Fixed Function Board</td><td>Fixed Function Board</td><td>Fixed Function Board</td></tr>
<tr><td colspan="3">Computing Platform</td></tr>
</table>

Each board in the system performs its own fixed function: fax, voice, etc. This made sense when DSPs were expensive and relatively slow, PC's were running the DOS operating system and computer memory was hard to come by. Today's DSPs are outpacing Moore's law in their increase in computing speed; PC's are running Windows NT and UNIX, and computer memory is inexpensive and readily available. In today's computing environment this traditional structure no longer makes sense.

Commetrex believes that the use of intelligent fax boards has been limited by their high per-port prices and complex installation procedures. DSPs are becoming so powerful, and the number of systems that are limited to one function, such as fax, are becoming far fewer. So fixed-function fax boards are giving way to powerful, integrated-media multiline resource boards like those that provide the hardware foundation for the M-Series.

Commetrex combines its MultiFax software, including fax modems and T.30, with programmable DSP-resource boards that can support multiple media. For the PowerFax M-Series they're configured as "fax boards". The diagram below shows what a system looks like when built from these value-adding components.

Application
Multi-Function System Resources • Fax • Voice • Data
Computing Platform

In 1993, using this technology, Commetrex originated the first software-defined fax resource and forever changed the industry with its innovative approach to creating board-level telecommunications products. In addition to intelligent fax boards, this same technology has been used by Commetrex and its technology partners to create a number of commercially available fax machines and multi-function peripherals.

The PowerFax M-Series

Commetrex "leveraged the efforts of others"in this case Natural MicroSystems (NMS) by taking NMS's DSP-resource boards and adding Commetrex's market-proven MultiFax fax software, to create a comprehensive product line including the analog M-4 and M-8, and the M-T1, the industry's only 24-port board with an integrated T-1 network interface. The M-Series has PTT approvals in 35 countries.

The PowerFax M-Series employs the latest in DSP technology to significantly reduce the cost of PC-based fax boards. The analog boards are available with loop start and, optionally, DID interfaces. The M-T1 provides a DSX-1 line interface with D3/D4 framing (ESF or SLC-96 available on special order).

The M-Series is suitable for information service systems, fax-server, fax-mail and fax store-and-forward applications from high-volume fax broadcast systems to enterprise-wide LAN fax servers and SOHO integrated voice/fax systems. PowerFax M-Series fax boards are being used in the most demanding applications to send and receive hundreds of thousands of faxes a day.

Easy To Install And Use

The M-Series is unique among fax boards in offering automated procedures to install the Developer and Runtime software. Simply put the configuration software in the PC and run the installation program.

The M-Series boards include an "enhanced-compliant" MVIP switch to support advanced applications that require inter-board connection and switching. They have a '386-class control processor and a DRAM to reduce host-PC overhead. The fax modems are implemented on open-architecture Texas Instruments DSPs, allowing simple software upgrades to take advantage of the latest M-Series software releases.

The M-Series can dynamically re-size and re-encode fax images to suit the capabilities of the receiving terminal. This means a B4 image can be dynamically converted to A4 size should the receiving terminal not support B4. Similarly, MR- or MMR-encoded images will be dynamically converted to MH should the application require it.

There are three main data structures used in the M-Series MultiFax Document Queues (MDQs), Fax Status structures, and System Parameters. MDQs, which are managed with 7 functions, contain documents to be sent and received. This permits documents to be assembled independently of the fax application, supporting client/server architectures. The Fax Status structures reflect the status of the current fax operation, including such indications as current page, current document, fax negotiation results, transfer rate, resolution, encoding, and error codes. System parameters hold information such as the default values for subscriber ID, country code, and manufacturer ID.

The M-Series API includes queue-management, status, send, and receive functions, plus the functions for dialing, call-control and parameter-management. The Developer's Kit includes source code of sample applications and several utilities to aid in the development process.

MultiFax Fax Description Language (FDL) macros support the creation of faxes from dynamically assembled elements. For example, a personalized fax letter including information obtained from a centralized database, a pre-scanned signature, the date, time, and even a per-call personalized paragraph, can be created and transmitted. Examples of FDL macros are:

@CONVERT:END	@CCONVERT:PADTOEND
@CONVERT:FILE	@CONVERT:PAGEBREAK
@CONVERT:FONT	@CONVERT:PAGESIZE
@CONVERT:MARGIN	@CONVERT:TEXT

These macros are inserted into a text file and converted using the MultiFax file- and format-conversion functions.The resulting file is then placed in the appropriate MDQ.

The M-Series also supports a variety of format conversions. The files produced by word processing, page layout, and graphics packages are generally not in the file format a fax transmits and receives.The PowerFax M-Series supports conversions from FDL, PCX, Bi-Level TIFF, TIFF-F, and text (including system outline fonts).

Fax System Developers are offered a comprehensive set of APIs to create their systems. And, since all of the M-Series boards use the same fax APIs, adding or replacing boards for system upgrades and expansions is straightforward. These APIs include:

AllocFaxRes	GetConvertedPages
AnswerFaxPoll	GetDocStatus
CheckTiff	GetFaxResStatus
EndConvert	GetDocStatus
ReceiveFax	GetVersion
CancelFax	ClosePort
OpenConvert	CreateQueue
FaxConvertDirect	QueueDoc
SendFax	QueueFree
ReleaseFaxRes	ReleaseFaxRes
GetAppPorts	GetParms
GetAppResource	ResetSendQueue
GetConvertedPages	SetPageSize

For more information visit: www.commetrex.com

PowerFax M-Series Features

Feature	M-4	M-8	M-T1
Nominal # ports	4	8	24
V.17 (14.4K) Xmit/Rec	4/4	8/5	24/24
PC requirements	ISA (full size)		
Host OS	Windows NT		
Fax Modems: 2400, 4800,	x	x	x
7200, 9600, 12,000, 14,400			
Network interfaces	LS, GS, DID		T-1
Caller ID	Caller ID		CLID
Dialed # ID	DID		ANI
Dialing: DTMF/pulse	x	x	x
Encoding: MR, MR, MMR	x	x	x
ECM	x	x	x
BFT	x	x	x
Conversions: TIFF, PCX, DCX	x	x	x
Fax Description Language	x	x	x

PowerFax M-Series Specifications

	M-4	M-8	M-T1
HOST REQUIREMENTS:	3 86/486/Pentium PC-AT with Windows NT		
CAPACITY:	4 ports per board. 48 ports per PC	8 ports per board 48 ports per PC	One T1(DSX-1) 24 ports per board. 96 ports per PC
FAX ENCODING:	MH, MR and MMR. Error Correction Mode (ECM)		
FAX MODEMS:	V.21 (300bps) for T.30 negotiation, V.27ter (2400/4800bps), V.29 (7200/9600bps), V.17/V.33 (14400/12000/9600/7200bps)		
FORMAT CONVERSIONS:	ASCII, PCX, TIFF-F, TIFF, OS/2 and Windows NT System Fonts, FDL		
BOARD WARRANTY:	7 years		

	M-4	M-8	M-T1
TELEPHONE INTERFACE			
Line Interface:	Loop Start (optional DID)	Loop Start (optional DID)	
Connector:	4 RJ-48	4 RJ-48	RJ-48C
Impedance:	600 Ohms nominal	600 Ohms nominal	
Ring Detection:	Min. Threshold 40 Vrms 17-33 Hz	Min. Threshold 40 Vrms 17-33 Hz	
Loop Current Range:	18-70 mA	18-70 mA	
DSX-1 TELEPHONY INTERFACE			
Line Interface:			Complete interface to one T1 trunk
Framing:			D3/D4 (ESF or SLC-96 available on special order)
Insertion and Detection:			ABCD bits
Zero bits:			Selectable B8ZS, jammed bit (ZCS) or no zero code suppression
Alarm Signal Capabilities:			Yellow, Red and Blue
Counts:			Bipolar violation, F(t) error and CRC error
Robbed Bit:			Selectable on a per channel basis
Loopback:			Per channel and overall under software control. Automatic remote loopback with CSU option
HOST INTERFACE			
Electrical:	PC/AT bus designed to IEEE P966 ISA		
Mechanical:	Board designed to meet IBM's specifications PC/AT Prototype Adapter Reference Manual #6361674		
Bus Speed:	4-12 MHz		
I/O Mapped Memory:	128 KB of on-board interface memory accessed at DMA rates via I/O string move		
I/O Addresses:	Switch select any of the 64 I/O addresses		
Interrupts:	Choice of 7 software configurable interrupt lines with all boards sharing only 1 interrupt line		
POWER REQUIREMENTS:	+5 V 1.75 Amps +12 V 70 mA -12 V 100 mA		2.5 Amps per board of PC/At power at +5 volts
TONE DIALING			
DTMF Digits:	0-9, *, # and ABCD per ITU-T Q.23 and Q.24		
Rate:	Programmable (10 digits/sec nominal) Wait-for-dial-tone capability		
Dialing Parameters:	Software controllable		

	M-4	M-8	M-T1
TONE DIALING (Continued)			
Dialing Amplitude:	Network compatible (configurable by country) programmable range -33 dBm to 1 dBm		
PULSE DIALING			
10 Digits:	0-9		
Pulsing Rate:	10 pulses/sec (nominal)		
Make/Break Ratio:	Network compatible (configurable by country) 40/60 nominal		
DTMF TONE DETECTION			
DTMF Digits:	0-9, *, #, ABCD		
Dynamic Range:	-47 dBm to 0 dBm per tone, programmable		
Tone Duration:	40 ms (minimum)		
Acceptable Twist:	10 dB		
Talk-off:	Exceeds all Q.24 standards including Mitel CM 7291 and Bellcore TR-TSY-000762 tests		
ON-BOARD PROCESSORS AND MEMORY			
DSPs:	2 Texas Instruments TMS320C51 DSPs at 40 MIPS each	2 Texas Instruments TMS320C51 DSPs at 40 MIPS each	6 Texas Instruments TMS320C51 DSPs at 50 MIPS each
Microprocessor:	25 MHz 80386SX		
MVIP DIGITAL SWITCHING	Enhanced-compliant MVIP interface provides total flexibility in connecting T1 channels to DSP resources, to other T1 channels, or to MVIP bus timeslots. Note that connections between T1 channels, or between T1 channels and other on-board resources, do not tie up MVIP bus timeslots. MVIP timeslots are only consumed when needed for inter-board connections.		
REGULATORY CERTIFICATION (USA AND CANADA)			
Telephone Network:	FCC Part 68 and DOC CS-03		
EMI:	FCC Part 15 Subpart J, Class A		
Safety:			NRTL Recognized to UL 1459 and CSA Recognized to Standard 225
STANDARDS COMPLIANCE			
Digital Multiplexer Requirements and Objectives:			AT&T Pub. 43802, July '82
Service Description and Interface Specifications:			AT&T TR 62411, ACCUNET T1.5
Carrier to Customer Installation DS1 Metallic Interface:			ANSI T1E1/88-001R1, Feb. '88
ENVIRONMENT			
Operating Temperature:	0°C to 50° C		
Storage Temperature:	-20° to 70° C		
Humidity:	5 to 80%, non-condensing		

Chapter 16

Dialogic Fax Hardware Approach and Product Line

Introduction

From Fax Hardware to Universal Everything

The market for computer-based fax is strong and growing. And along with growth comes change. Today's fax technologies, like voice technologies before them, are quickly migrating to more modular and open hardware. Tomorrow's fax systems will be "universal everything" systems—just one part of the same unified, well-managed network that also handles voice communications. Instead of depending on expensive, dedicated (and dead-end) hardware, next-generation fax technologies will rely on improvements to software. The result will be tremendous savings in PBX systems, in management resources, and in the high price companies now pay to transmit fax over standard telephone lines.

Fax Reaches Critical Mass

Fax systems are no longer an option. Over the last decade, fax has matured into an essential tool for doing business, as common as lights or telephones. Market studies show that fax usage has grown at a breakneck pace that won't soon slow down. The number of pages faxed each year is expected to have doubled from about 350 million in 1996 to more than 700 million worldwide by 2000. (Source: IDC, 1997)

Although fax usage is growing steadily worldwide, the growth is most dramatic in Asia. By 2000, Asia will fax some 325 million pages per year—nearly half the worldwide total. (Source: IDC: 1997) Cultural considerations are one reason for the tremendous popularity of fax in Asia. Although they are complex, with numerous regional and cross-country dialects, Asian languages are much more graphic that European languages—making written communication via fax an efficient alternative to voice communication. Sending fax communications can also help minimize the difficulties of communicating across time zones, another consideration in Asian markets.

In Asia, as in the rest of the world, all this fax usage adds up to big business. The cost of the fax systems themselves is only the beginning. The real expense behind most fax transmission is the cost of using telephone lines—often at the highest per-minute rates of the day. The average midsized company spends an astounding $37 million each year on telephone expenses, with about 41% of that expense dedicated to the cost of fax transmission. (Source: Pitney Bowes Gallup Study, 1996.)

Market statistics also show that companies are moving away from fax servers and toward software-based solutions. The market for Windows NT fax server software grew tenfold between 1995 and 1997, in both revenue and the number of ports in use. (Source: IDC, 1997.) Today, only about 20% of fax transmissions are generated from the desktop, with the rest coming from shared fax machines. Instead of having individual users fax from their desks, companies are finding it makes sense to use their fax servers to manage their overall fax traffic.

The World of "Universal Everything"

The progression to software-based fax technologies is a natural one. In a world where technologies come and go and a company's communication system develops over a long period of time, there comes a point when a company must decide how to turn a collection of diverse, often incompatible technologies and equipment into a single, unified communications system that maximizes existing resources and optimizes efficiency and costs.

Today, companies are finding themselves with an inefficient collection of separate, self-contained fax systems that must vie with equally self-contained voice mail systems for space on the same PBX. For example, a company taking a fresh look its PBX resources might find significant periods during the day when the lines dedicated to fax are not being used. At the same time, the lines dedicated to voice communications are probably quiet all night. When the company needs to expand, it has two choices: add expensive PBX resources, or find a way to get more use from the resources it already has—by managing voice and fax on the same network and sending fax transmissions when the voice lines are quiet.

The key to this transition is fax technology that runs on modular, open hardware instead of the closed, proprietary hardware that defined—and constrained—yesterday's fax systems. The transition to open hardware has already taken place in computer telephony, where standards are well established. The result has been tremendous user benefits, including lower costs and wider flexibility to combine features and technologies from different companies.

A Revolution in Fax: A Revolution in Computer Telephony

To meet these emerging challenges, Dialogic, the leader in computer telephony (CT) platforms, has designed the DM3 mediastream architecture for open call processing solutions. The DM3 architecture is a leap forward in open platform flexibility, price, performance, resource integration, and density. The DM3 architecture supports multiple firmware resources from multiple vendors, and offers the DMFast TM development environment for the rapid development and integration of call and media processing resources. The DM3 architecture also enables solution developers to build industry-standard, interoperable solutions based on Signal Computing System Architecture TM (SCSA) compatible software while protecting their underlying hardware and software investment.

The DM3 Design

The DM3 architecture consists of embedded software modules and hardware that provide a platform for developing leading-edge call processing applications. DM3 supported platforms are modular, scalable hardware implementations of the architecture and currently include high-density PCI, CompactPCI, and VME platforms. Other platforms including low-density PCI will also be available. Platforms are integrated with firmware resources to create product bundles.

Software Architecture

The DM3 software model consists of host software that runs on a PCI, CompactPCI, or VME target system and embedded software that runs directly on the DM3 platform. This layered and modular software design includes a host device driver, host interface library, Kernel and underlying real-time operating systems, and technology resources.

Software Architecture Overview

The DM3 Kernel is embedded software that is common to all processors on the DM3 architecture. The real-time operating systems (RTOs) used beneath the Kernel include SPOX (for DSPs, such as the Motorola 5630x) and VxWorks (for RISC, such as the i960 control processor and PowerPC 603e signal processors). Using industry-standard operating systems and tools helps reduce the time it takes to port firmware resources and develop solutions. The DM3 Kernel also offers runtime services for both the CP and SPs, and maintains a communication path between the DM3 firmware resources and the host software. The Kernel provides a message sending mechanism and timer services, plus resource, configuration, and memory management services.

DM3 firmware resources are embedded algorithms that run on the hardware using the Kernel services. Resources can be divided into processor-specific applications, such as a player resource containing various types of decoder components. Components are assigned unique DM3 addresses and communicate with one another via messages that are routed by the DM3 Kernel.

On the host software side, DM3 driver support is a technology-independent transport layer linked to a DM3 host interface library. The driver device passes messages and data from the host to the resources on the appropriate DM3 board using the functions of the Kernel.

Host software libraries, such as a voice library, communicate with firmware resources through the DM3 host interface library. Host software libraries are shielded from the device driver by the DM3 host interface library, providing protection from differences in host operating systems. To ease communication between the host and firmware resources, the host interface library enables driver configuration, component management, message transfer, and data read/write operations.

Firmware resource modules perform the following types of functions:

- Call control
- Tone generation and detection
- Call progress monitoring
- Voice record and play
- Fax
- ASR
- TTS generation (speech synthesis)
- Switching, bridging, and routing
- Audio conferencing
- IP voice and fax

The DM3 architecture offers high-level API support based on the Enterprise Computer Telephony Forum (ECTF) APIs, as well as lower-level support based on the DM3 Direct Interface. DM3 products are also forward compatible with existing Dialogic network interface, voice, and fax APIs. All of these interfaces are open and portable across CompactPCI, PCI, and VME hardware platforms.

Firmware resource developers can write their own algorithms to the DM3 Kernel, which helps to insulate the developer from the low-level details of the real-time operating system and increases processor independence, code portability, and supportability.

High-Density Platforms

The DM3 high-density platforms consist of the following:

- A DM3 Architecture baseboard (CompactPCI, PCI, and VME form factors)
- A network interface daughterboard with 1 to four T-1 or E-1 ports
- Or a different network interface daughterboard that provides four or eight ports of analog (TBR 21) or BRI network interfaces
- Up to three stackable signal processor (SP) daughterboards supporting up to six processors each

The DM3 high-density platforms consist of three baseboard types: PCI, CompactPCI, and VME. These baseboard types support daughterboards for signal processing and computing, communications for development, and additional network interfaces (T-1, E-1, or ISDN).

The DM3 high-density baseboard contains an Intel i960CF control processor (CP), associated RAM and flash memory, two complete network interfaces, two SC4000 ASICs, configurable global memory, host RAM for communication between the CP and the host, and a mediastream management ASIC (the MMA).

The MMA transfers all bulk data between the host RAM, CP, SPs, and SC4000 ASICs via a 32-bit DMA (direct memory access) bus to and from the DM3 global memory. (The global memory is installed on SIMM or SO DIMM modules for easy upgradeability and is available in different sizes, up to 16 MB.) The MMA also commands and moves PCM data between the SC4000 ASICs and the various processors' memories via the PCM buffer.

The single CP on the DM3 baseboard manages SCbus access and controls the SC message bus via the MMA and SC4000 ASICs. Data on the SCbus is available in multiples of 64 Kb/s TDM channels (bundles), in linear format or encoded in A-law or (-law format. The two SC4000 ASICs can access up to 256 of the 2,048 SCbus time slots, allowing up to four T-1, E-1, or ISDN trunks with echo-canceled data that can be exported to other technology resources, such as ASR.

Each SP daughterboard is configured with up to six Motorola 5630x DSPs or up to four PowerPC 603e processors. Other signal processors can also be used on SP daughterboards, and daughterboards with different processors can be combined for maximum flexibility. The same SP daughterboards can be used on PCI, CompactPCI, and VME baseboards. The communications daughterboard is used for debugging or direct access to the board during development. The digital network interface (DNI) daughterboard provides two additional T-1, E-1, or ISDN network interfaces to supplement the two network interfaces on the baseboard. Note that the SP daughterboards are stacked in a different location on the baseboard from the other daughterboards. This allows up to three SP daughterboards, a communications daughterboard, and a DNI daughterboard to be stacked on a single baseboard. A baseboard with a single SP daughterboard requires only one PCI, CompactPCI, or VME slot. A baseboard with two or three stacked SP daughterboards requires the space of two slots.

From Architecture...To Products

The DM3 Fax Series will include the features of the industry-leading Dialogic GammaLink CP product line, including GammaLink quality T.30, ECM, MMR compression, and the proven ability to connect to more fax devices than any other comparable devices. It will mirror the GammaLink product line, with DM3-based products equivalent to the company's existing products.

With industry-standard PCI bus expansion boards and a variety of channel densities to choose from, customers will be able to integrate industry-standard voice products easily into exactly the system they need, optimizing both price and performance. Customers can add features and grow their systems while protecting their investment in hardware and application code. Since the DM3 fax boards operate in SCbus configurations, if customers wish to add resources such as automatic speech recognition (ASR), voice play and record, or text-to-speech (TTS), the resources can be switched via the SCbus.

DMFax Product Family

DM/F 240, DM/F300 Fax Only Resource

The Fax Resource boards are primarily designed to replace the existing CP12 based GammaLink fax product line. The fax resource products are ideal for voice centric applications, or solutions that require a smaller number of fax ports to a larger number of voice ports. This is typically in CT Server or Telco-type environments.

The DM/F 240 and DM/F 300 are 24 and 30 port (respectively) fax resource products with the full Dialogic fax feature set. The products are available in Windows NT on the PCI, cPCI form factor, and on Solaris in the PCI and VME form factors. A DM/F 480 and DM/F 600 is available upon request (DSP and firmware upgrade).

DM/F240 T1, DM/F 300 E1: Fax and Network Interface

This is the ideal product for traditional fax applications such as LAN fax and fax broadcast. The DM/F 240 - T1 and DM/F 30 E1 provide a single slot, full span of fax processing with integrated T1 or E1 network interface. The product provides the highest density integrated solution on the market today. A 48 and 60 port version is scheduled for release in Q3 1998.

The DM/F 40 is a four port analog version of the same integrated technology. This product provides identical programmatic interface to the T1 and E1 solutions allows for a truly scalable product family from 4 ports 60 or more.

DM/VF 240 T1, DM/FV 300 E1: Voice, Fax and Network Interface

The ultimate goal is the integration of the voice, fax and network interface technologies onto a single platform, this product is the DM/VF 240-T1 and DM/VF 300-E1. There is a DM/VF 40 version as well.

Key Benefits

More than a technology platform, DM3 is a whole new approach to open CT. The DM3 modular architecture takes advantage of both the SCSA software and hardware models, making it ideal for multiple-resource applications from multiple vendors. DM3 customers can compete on a new playing field with lower hardware, development, integration, and support costs, which makes it possible to continually tap into new markets by creating applications that run across all DM3 platforms.

Unparalleled Flexibility

With its hardware independence, the DM3 platforms offer unparalleled flexibility. For example, a simple firmware load is all that is needed to transform a system from one application (such as a call center) to another application (such as a broadcast fax server). No hardware changes are needed except for adding more processing MIPS when appropriate.

If a new application requires an increase or decrease in memory or signal processing resources, memory modules and signal processing daughterboards can be added or subtracted as needed. The same signal processing daughterboards are compatible among CompactPCI, PCI, and VME baseboards.

Breakthrough Performance

With up to four T-1, E-1, or ISDN ports per slot, the DM3 architecture offers a breakthrough in performance and density. Technology resources can be optimized for improved performance by writing to the DM3 Kernel API, instead of writing to a specific signal processor. Therefore, firmware resources can be partitioned between the control processor that runs on the DM3 baseboard (for control and management functions), the DSP that runs on the signal processing daughterboard (for fixed-point signal processing functions), and the RISC processor that runs on the signal processing daughterboard (for floating-point signal processing and RISC processing functions).

Price/Performance Leadership

The DM3 platform is built to maintain its price/performance leadership. For example, the high-density platform avoids complex arbitration mechanisms and has a vastly simplified timing scheme between processors and memory, allowing for easier upgrades to faster and more powerful processors and memory. Furthermore, a special high-speed Dialogic ASIC manages all data transfers on the board, freeing the control processor for message and task management. As Dialogic delivers products faster, customers are able to keep pace and stay on the leading edge of the price/performance curve.

In addition, technology resources are written to the DM3 Kernel instead of being tied to a specific signal processor or its embedded real-time operating system. This allows firmware developers to take advantage of ongoing improvements in signal processing technologies, keeping their firmware resources on the leading edge of the signal processing performance curve.

The Wide-Open Future

Fax has become an essential business tool. And tomorrow's fax systems will deliver new features for a lower cost as new systems are built on open, standard hardware. The move to open standards will let users optimize their voice and fax transmission over a single network and using the Internet and corporate intranets to transmit without the high charges they are paying today.

For more information, visit: www.dialogic.com

Chapter 17

The Natural MicroSystems Hardware Approach and Product Line

Introduction

Natural MicroSystem™ Corporation, founded in 1983, is a leading provider of open, scalable computer telephony platforms, or Open Telecommunication™ solutions. Their state-of-the-art technology enables a growing international network of partners to dramatically shorten time to market for innovative, high-value computer telephony systems and applications, including Internet voice and fax systems; network-based enhanced services; IP telephony solutions; and telecommunications infrastructure applications.

The value of Open Telecommunications to developers is the ability to take advantage of standards-based building blocks, reducing time to market when creating complex applications. The Natural MicroSystems family of products provides fundamental functions that adhere to widely-accepted standards for easy integration into communications networks around the world. Our enabling technology works with standard computing platforms, such as PCs, so that developers can take full advantage of the wealth of products, tools, and support that open systems provide.

Fax Technology Overview

Natural MicroSystems supports fax processing via NaturalFax™, the industry's highest-density solution for high-value fax service applications. NaturalFax, a member of the Natural Media family, enables service providers to reduce costs by offering fax services worldwide without additional dedicated fax modem boards. Host processing is minimized permitting up to 30 ports of fax in a single PC slot and 120 fax ports on an Intel Pentium 200. Through integration with Natural MicroSystems' CT Access API, NaturalFax allows developers to enhance their existing voice applications and capitalize on the growing demand for fax service applications, such as fax-on-demand, Internet fax, and fax broadcast.

The Natural MicroSystems NaturalFax platform consists of NaturalFax software, an Alliance Generation® digital signal processor (DSP) resource board, and the CT Access development environment. Major components include the NaturalFax API and supporting managers which handle file-format conversions, file services, Group 3 fax negotiation, and image transfer. The API provides developers with NaturalFax queuing, status, send, and receive functions, plus the CT Access functions for voice play/record, call management, parameter management, and database access.

NaturalFax, through integration with the CT Access API, enables developers to add fax communications to other media support—such as voice processing and text-to-speech—on the same PC board. In addition, the call control management function of CT Access and NaturalFax enables switching between voice and fax functionality automatically on a port-by-port basis. For example, upon detection of a fax calling tone, the port will automatically switch from a voice port to process the incoming fax document. When document processing is finished, the port can revert to its usual voice port status. This flexibility enables developers to maximize performance without the need to dedicate specific ports to one function.

NaturalFax can support applications that transmit and receive Group 3 facsimiles at rates of 2400, 4800, 7200 or 9600 bps. NaturalFax can also transmit at 14,400 and 12,000 bps. NaturalFax supports polling remote fax terminals, and its software licensing scheme requires no additional hardware security devices.

The Hardware Environment

NaturalFax runs on Alliance Generation (AG) boards without any additional hardware. The Alliance Generation (AG) product line is part of the Natural Platform family. Natural Platforms hardware and firmware products, with their state-of-the-art design and global approvals, deliver maximum performance, the highest capacity in the industry, and unparalleled flexibility for the development of Open Telecommunications solutions. Natural Platforms offer a large number of solutions, all built on the open, industry-standard MVIP and H.100 architecture, enabling access to technology from multiple vendors for true interoperability.

The Alliance Generation Family

The award-winning Alliance Generation product line is a family of full-function DSP boards and an integrated software environment for the development of systems that require voice and call processing, fax, switching, and integration of telephone systems and computer or database systems. Each AG board can support a daughterboard which can either expand the number of ports or add support for additional functions, such as speech recognition. The Alliance Generation architecture uses DSP resource capabilities efficiently, allowing it to go beyond available technology in terms of port density, functional performance, and the integration of multiple technologies.

Features of the AG family include:

- On-board resources that reduce host overhead and allow for high-capacity call processing
- Dynamic, efficient task processing
- Advanced solutions that provide from eight to sixty ports in a single PC slot
- MVIP support, ensuring open architecture and vendor independence
- Capability for a wide range of system sizes and configurations

The AG family includes DSP resource and network interface boards, and DSP resource boards. The following AG family boards support NaturalFax: AG-8/80, AG-8/80 DSP, AG-T1, AG-E1, AG-24+, AG-30, AG-48, AG-60, AG-T1, and AG-E1.

Resource and Network Interface Boards [AG-8/80, AG-T1, AG-E1]

The resource and network interface boards support both voice processing and network interfaces. The AG-8/80, with 80 MIPS of DSP power on-board, provides 8 ports of voice processing along with analog telephone connections for loopstart or DID interfaces. The AG-T1and AG-E1 platforms contain 24 and 30 ports of fax and voice processing, respectively, and an integrated digital trunk interface. As many as 30 fax modems can execute on a single AG-E1, giving the AG-E1 the highest fax density in the industry incorporating an on-board network interface. These resource and network interface boards are designed to plug into a PC's ISA bus.

Resource Boards [AG-24, AG-30, AG-48, AG-60, AG-DSP/80]

The AG resource platforms are designed for use with separate network interface boards. The AG-24 provides 24 ports of voice and call processing and is upgradable to 48 ports. The AG-30 provides 30 ports of voice and call processing and is upgradable to 60 ports. The AG-48 and AG-60 provide 48 and 60 ports of voice and call processing while preserving a single PC slot solution. These boards will also support up to 48 ports of fax without a network interface in one ISA slot. The AG-8/80 is also available in a DSP resource version (AG-DSP/80).

Alliance Generation²

Digital Signal Processors (DSP), the specialized microprocessor chips that are the core of scalable, open telecommunications servers are, like computer chips, constantly growing in power. Natural MicroSystems is developing the Alliance Generation² which is designed to take advantage of the next generation of DSPs, the C54x family from Texas Instruments. The Alliance Generation² family is an evolution of Natural MicroSystems' AG family that offers developers dramatically faster processing, increased scalability, and easier customization. Alliance Generation² is fully compatible with the current Alliance Generation architecture, so existing software applications, written using CT Access software, can take full advantage of increased performance without requiring any code modifications.

With the more powerful DSPs, each port on an Alliance Generation² has more MIPS (millions of instructions per second) available, enabling more robust applications. The encoding and decoding of media such as fax requires a large number of MIPS. The Alliance Generation²'s substantial increase in MIPS per board and MIPs per chassis ensures more powerful applications and lower cost platforms.

The Software Environment

NaturalFax is a member of the Natural MicroSystems Natural Media product family. It is also tightly integrated with CT Access, the Natural Access software development environment.

Natural Media

Natural MicroSystems' Natural Media products extend the Telecommunications Services Architecture (TSA) beyond voice and call processing to include support for mixed-media, such as fax, speech recognition, and text-to-speech. Natural Media integrates these additional media seamlessly into a single unified set of capabilities for the application. This results in reduced application development time for complex mixed-media applications, and significantly reduces production costs.

NaturalFax

NaturalFax software, when combined with Alliance Generation hardware, enables applications to send or receive standard fax communications from files on the computer. NaturalFax demonstrates the unique flexibility of the Alliance Generation architecture, as no additional hardware is required. Fax modem functionality is fully integrated with voice and tone processing in the AG DSP platforms to create universal ports. There are three significant benefits to the universal port:

- **High efficiency** - Fax capability is added to the application increasing the application value without consuming backplane slots or system power. The fax modem is implemented as DSP software and shares the same resources as the voice/tone functions.
- **Simplicity** - Development time is reduced when fax capability is available to every port in the system without switching. Utilizing fax is as common and natural as voice and tones.

- **Efficiency** - As many as 24 fax modems can execute on a single AG-T1 and 30 ports of fax on the AG-E1, giving the AG boards the highest fax density in the industry.

NaturalRecognition and NaturalText

NaturalRecognition supports speech recognition with software and the addition of a daughterboard for the Alliance Generation platform. It enables application programs to accept spoken input using automatic speech recognition (ASR). NaturalRecognition provides both speaker-independent and speaker-dependent support with vocabularies available off-the-shelve for dozens of languages.

NaturalText supports text-to-speech (TTS) technology on additional daughterboards. TTS is ideal for the emerging unified messaging market which exemplifies the technology convergence in TSA. Unified messaging provides professionals with simple and consistent access to multimedia messaging such as email, voice mail, and fax. Each of these media is transportable over LANs and WANs, so that multimedia correspondence can be conducted via computers. The telephone allows the message recipient to access this desktop data from any place at any time.

TSA unifies and simplifies the development of applications based on these technologies. On a single AG platform, the system developer has telephone network connectivity, voice and tone processing for prompts, fax, speech recognition, or text-to-speech. While a single board is the most efficient and simplest development alternative available, more complex multi-board and multi-chassis applications are also supported.

Natural Access

Natural Access is comprised of development and run-time software supporting Natural Platforms, Natural Media, and third-party products. The Natural Access software development environment provides a single, unified API for creating applications that integrate multi-media functions such as fax, speech recognition, and text-to-speech. Natural Access products simplify Computer Telephony application development and greatly improve system efficiency by executing these multi-media functions, as well as call control and standard voice processing, on a single board.

The cornerstone of the Natural Access products is CT Access. CT Access is available on Windows NT, SCO UnixWare, SCO OpenServer, Solaris, and OS/2. It supports an API rich enough to build applications rapidly and powerful enough to build specialized custom features.

CT Access

CT Access is an integrated software product for mixed-media telephony application development that supports the following:

- Natural Call Control™, a full-featured and flexible call control architecture
- voice buffering for play and record functions
- DTMF tone detection, queuing, and generation
- programmable tone detection and generation using DSPs
- other low-level functions such as signaling bit functions and 1200 baud FSK modems
- a full-featured voice file system that simplifies playing and recording voice messages and computed strings such as dates and monetary units.
- switching for ECTF H.100 and MVIP TDM buses
- NaturalFax which uses a software-based modem that shares the same DSPs as other CT Access voice and tone functions
- NaturalRecognition - ASR
- NaturalText text-to-speech

NaturalFax is a C function library component of CT Access that provides fax functionality.

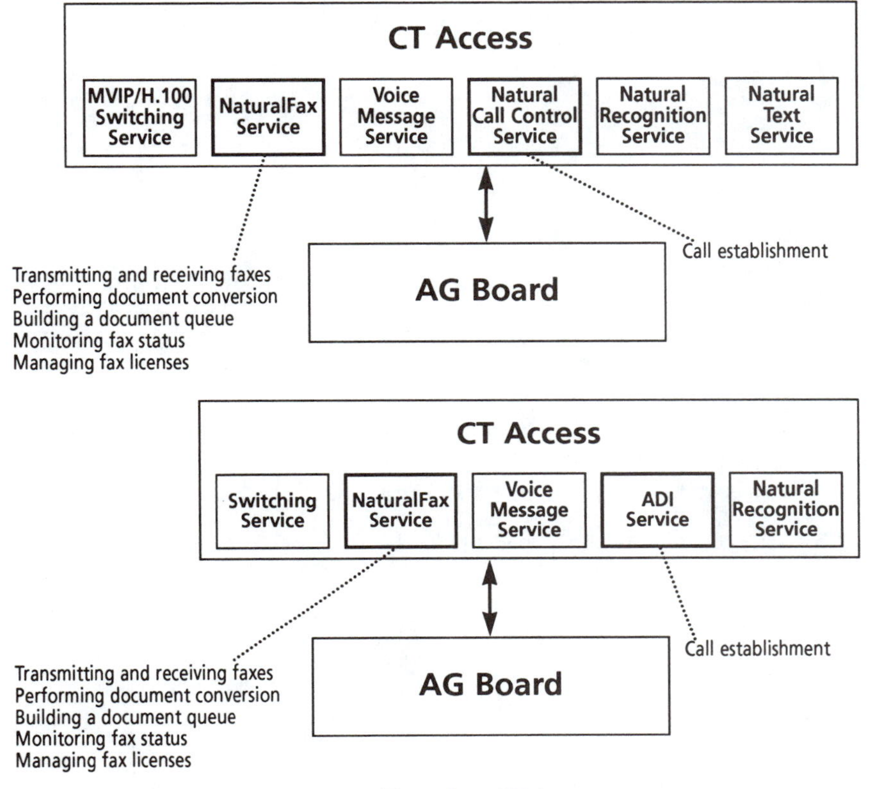

NaturalFax for CT Access

Conclusion

Fax is a vital component of the rapidly growing field of telecommunications. The industry's technology requirements are rapidly changing, constantly reaching new levels of sophistication as the demand for more communication features and volume grows ever stronger. As this technology evolves, fax, voice, video, and data are converging to provide more efficient and cost-effective methods of transporting information. The Telecommunications Services Architecture provides the highest scalability fax platforms, and the most integrated environment for the support of complex, high-value fax applications. This powerful combination gives solution providers a significant advantage in today's marketplace. For more information visit: www.nmss.com.

Section VI

Internet Fax

Chapter 18

Brooktrout Technology Internet Fax Strategy

The world of telephony has been a driving force behind and beneficiary of digital circuit and computer technology. More recently, digital data network technology has been applied to telephony applications including both voice and fax transmission. Voice and fax traffic now can be transported over packet data networks using the Internet Protocol (IP). This IP telephony is being used to reduce the cost of deploying and operating traditional voice and fax services, to create new services such as PC to phone applications, and to reduce the cost of network ownership by consolidating voice and data networks.

Brooktrout has been producing technology components for service providers and OEM system manufacturers for many years. These components are used by Brooktrout's customers to create voice and fax messaging systems. Brooktrout recently announced a new line of products for development of standards-based voice and fax services over IP packet data networks. These products are based on the Brooktrout Open System Telephony (BOSTON) architecture, a software architecture that provides developers with a unified Universal Port(development environment that speeds product development and provides a new family of low-cost, high-density system components.

Of the two telephony services, voice and fax, fax was the first to be transported over data networks. Today, several companies provide low cost fax services over packet data networks. Initially these services were provided as store-and -forward fax only over X.25 packet data networks. Today, both store-and-forward as well as real-time fax can be provided over widely deployed IP data networks.

Store-and-Forward Service – Store-and-forward fax service is a messaging technology. A fax is received over the PSTN by the IP fax gateway. The store-and-forward gateway acts as a receiving fax machine and stores the fax at the gateway. The fax is then converted to a series of messages and is sent to the receiving gateway over an IP data network. The receiving gateway then acts as a fax machine and sends the fax to the receiving fax machine over the PSTN. When the remote IP fax gateway receives confirmation from the receiving fax machine, it sends a confirmation message back to the originating gateway which then notifies the sender.

Store-and-forward fax has the following advantages:

- Enhanced functions such as auto retry, scheduled delivery, and broadcast fax are easy to add
- It is easily integrated with email systems
- It can utilize excess network bandwidth by deferring delivery to off- peak traffic times

A typical IP fax gateway is shown in Figure 1. This gateway is an IP fax server built with a standard digital telephony interface board (T1/E1 or PRI), and standard digital fax boards in a PC. The telephony interface and the fax boards are interconnected by either an MVIP or SC TDM data bus. Standard telephony interface and fax boards are used because they have international approvals and have been integrated with a large variety of PBX systems.

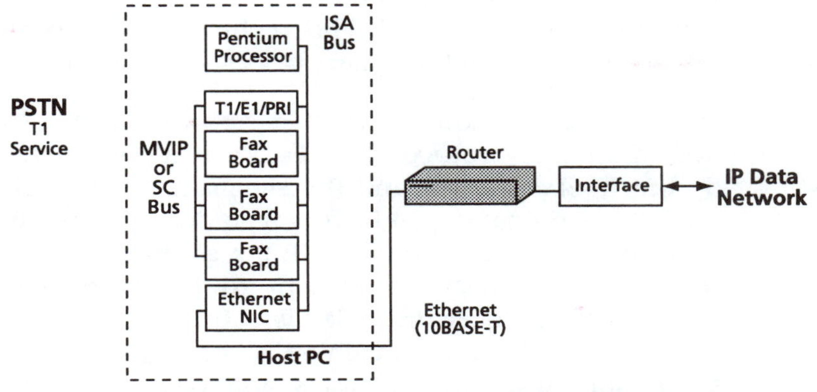

Figure 1. Typical Store and Forward IP Fax Gateway

Real-Time Fax Service

Although store and forward fax is a proven technology, it has the disadvantage of delayed confirmation of delivery. For this reason, fax technologists have developed real-time fax transport methods that provide an immediate delivery of the fax and an immediate confirmation of that delivery. Real-time IP fax is transparent to the fax machine users. The fax machine behaves exactly as if it were connected through a PSTN link.

FaxPAD and FaxRelay - There are two popular methods for implementing real-time fax over IP networks-FaxPAD and FaxRelay. Each method has its advantages.

- **FaxPAD** is a method that involves the manipulation of the T.30 fax transport protocol (the fundamental protocol of fax communications) to keep the sending and receiving fax machines operating even when data network delays are long. The T.30 protocol specifies time limits for certain communications during the fax transmission. If these time limits are exceeded, normal fax machines "time-out" and the transmission fails. In FaxPAD, when a fax machine "time-out" is likely due to long delays in the network, the local FaxPAD gateway employs different schemes to "spoof" the fax machine, thwarting the "time-out" and keeping it on-line until the next message is received. This technique is particularly effective on long delay IP networks such as the Internet.

- **FaxRelay** is a method that does not involve manipulation of the T.30 protocol. With FaxRelay, a fax is detected by the IP fax gateway and the signals are demodulated into their original digital format. The demodulated fax is then relayed to the receiving IP fax gateway where the signal is re-modulated and sent to the receiving fax machine. If the network delay is short (less than 1 second), the receiving fax machine accepts the signals and faxes normally. The gateways and data network are transparent to both fax machines.

Where to use FaxPAD or FaxRelay methods depends largely on the delay of the data network. The diagram in Figure 2 shows that networks with delays of one second or less can use the FaxRelay method. Networks with longer delays (up to three to six seconds), will perform better using FaxPAD.

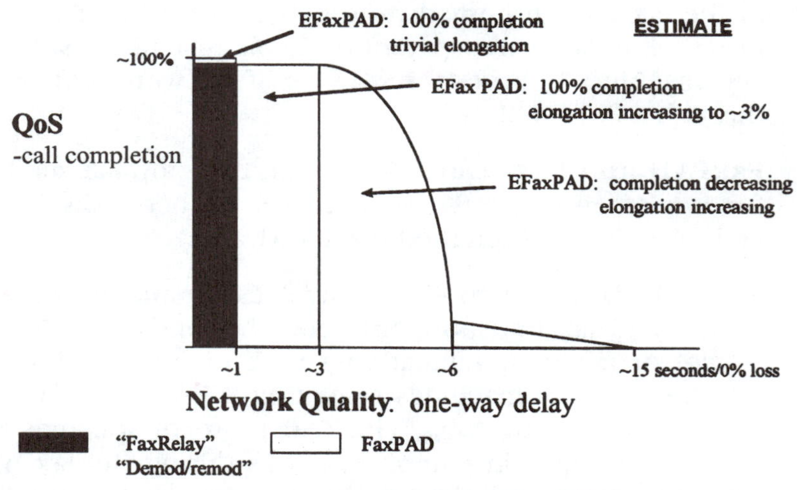

Figure 2. Real-Time IP Fax Options: FaxPAD and FaxRelay

IP Fax Standards

As IP fax becomes more generally deployed, standards are becoming important to provide interoperability between systems. IP fax standards work has been progressing steadily for a number of years now.

- **Store-and-forward fax** - The IETF (Internet Engineering Task Force) has been working on an addendum to the SMTP (Simple Mail Transport Protocol) which involves a MIME attachment for fax. This attachment uses the TIFF-F format for normal fax and TIFF-FX for more advanced fax features. The SMTP standard integrates fax with email.
- **Real-time fax** - These standards are being developed by the ITU' s TR-29 Committee. The X.38 standard for FaxPAD over X.25 networks has already been approved and extensions are being developed for FaxPAD over IP networks. The ITU and IETF are working together to ensure that these two standards don't conflict or duplicate each other.

For a discussion of Brooktrout's IP Fax Products, see Chapter 26, or visit: www.brooktrout.com

Chapter 19

Dialogic Internet Fax Hardware Strategy

Introduction

Internet fax. Many have said it's the most exciting thing to happen in the fax world in the last few years. There is as much hype as promised revenues. In an effort to explain Internet fax, or more correctly fax over packet networks, it has become apparent there is no "one size fits all" strategy. Different technologies and different protocols are required depending on the application, quality of service required, and most importantly, quality and integrity of the network.

Dialogic Corporation has marketed and released products to support the Dynamic Realtime™ method of sending fax traffic across the Internet. Dynamic Realtime has proved to be an excellent method for transporting fax data over unreliable networks such as the Internet. However, as with all processes, there are trade-offs between overhead and throughput.

We intend to discuss not only our vision and strategy for sending faxes in realtime over the Internet, but other types of packet data networks. We will investigate different schemes and their relative merits, and finally present product plans to address each of these markets.

The future of Internet Telephony, and Internet fax specifically, at Dialogic revolves solely on the new DM3 architecture. DM3 is very high powered, scalable DSP platform capable of providing upgrade-able technologies and services. See the section on "Hardware Platform" for more information.

Fax Markets

Until the last few years Computer Based Fax has only been used for "fax messaging." Fax messaging is the storing and forwarding of documents to be faxed. Fax Broadcast is exclusively a messaging technology. A fax transaction is forwarded 10,000 times to be sent out. The sender is notified after the transactions complete that each individual job was successful.

LANFax is a fax messaging technology. Documents or emails are forwarded to a LANFax server to be delivered when possible. The user is notified after the transaction completes that it was successful.

However, fax machine traffic, or fax machines sending to fax machines via the PSTN, is a real time operation, or a session based transaction. A user can put a document into a fax machine, send the document, and see a confirmation that the transaction completed successfully.

Messaging based transactions are on one extreme, session based transactions are the other. The world of Internet fax is a blending of messaging and session based solutions tailored to provide a solution that is ideal for the application. Different types of Internet fax solutions need to balance the following characteristics:

- the quality of service required by users of the system
- the quality of the packet data network

Market Segments

There are several different market segments for Internet fax, and Internet telephony in general. The markets can be easily divided into messaging and session based solutions.

Messaging based Internet fax

Because the following applications can be developed using the standard fax product set, we will simply describe the different technologies.

The first is email to fax. This is quite simple and quite pervasive. Many companies are providing an SMTP based Internet fax server. The basic concept is to provide email a message to a fax gateway located somewhere in the world which will fax your email message as a local call.

The next popular application is to connect multiple LANFax servers using the Internet or intranet as the backbone. This allows distributed corporate sites to route fax transactions to the least cost node for final delivery.

Session based Internet fax

The overwhelming theme in session based fax on data networks is to bypass the long distance toll charges. However, there are other issues which will undoubtedly drive this marketplace. First is a fundamental shift from investing in voice/TDM networks which are basically single function to a data network infrastructure capable of voice/TDM as well as video and traditional data.

Large corporations and telephone companies are realizing that investment in data communications infrastructure will serve them better in the long term providing a channel to carry voice, fax, video simply as the standard data packets already used. Single investment, single management and better utilization of the installed equipment.

Replacing the PSTN

This not going to happen any time soon. It is a widely held believe the Internet telephony will not replace the phone network. It will augment the existing services provided by the PSTN. Additionally, significant investment in the general purpose data network will be a tremendous growth opportunity in the foreseeable future.

The idea behind Internet telephony is to enhance the services provided by the PSTN, and provide new services not realized as yet.

The key players in this new market opportunity are traditional telecom suppliers such as NorTel, Lucent, Ericsson selling "edge devices" to companies like ATT, MCI, Sprint, WorldComm and other teleco's around the world. The market potential is viewed as anywhere from 100,000 ports up to tens of millions of ports. This is the market to expand traditional phone service with data networks.

The other "side" of this market potential is the Internet Service Providers. The key to Internet telephony in markets around the world is access to POTS lines and to reliable pipes into the world's IP network. ISPs have both. And again, an edge device is ideal for this market as well.

Edge devices are defined as black boxes which plug in TDM/PSTN trunks and connect into a data network. Edge devices generally provide conversion from TDM to IP and back. These boxes must handle the different types of traffic managed over the PSTN. Therefore, they must discriminate between voice, fax and potentially analog data calls, and provide "appropriate" methods for converting the traffic to the packet data network.

Because data traffic over the PSTN is probably less then 2% of the total traffic carried, an appropriate method for managing that traffic would be to not route it over the data network, but to keep it on the PSTN.

Because the market here is to augment existing PSTN traffic on data networks, the nature and characteristics of the products and services provided must be "telco" quality. The data networks are typically very well managed, highly redundant and very high throughput. Therefore, it is possible to provide some realistic boundaries to the expectations of the network. It should be possible to define minimum bit error rates, maximum jitter values, average and maximum round trip latencies, as well as to provide a managed priority routing scheme.

This is in stark contrast to the wide open world of the Internet. While there are many independent studies of the characteristics of the Internet as a network, there is only one thing for sure, it is totally unpredictable.

Gateways

Gateways are Customer Premise Equipment (CPE) products which are collect voice and fax traffic and place it onto packet networks. Such products are already well established using Frame Relay as the network to connect field and remote offices together. The use of Frame Relay implies a managed network with definable characteristics. The obvious negative to such solutions is the need for different telecom equipment and different telecom services other than the data network which is presumably already installed and configured (for Internet access).

There appears to be an emerging market to provide these gateway devices to connect over IP, either in intra or inter net configurations. Internet adds an enhanced level of security requirement. However, this can be easily overcome with off the shelf encryption schemes.

Data Networks

There are many different types of networks, and this is certainly not an exhaustive list of the different types. The following discussion is intended to discuss the different aspects of networks which may be used for moving fax traffic over data connections.

There are four dimensions to the data network:

1. **Bandwidth** - defined here as the number of bits per second available to transport. Cellular networks provide bandwidth of roughly 14,000 bits per second, whereas a TDM network provides 64,000 bps.

2. **Delay** - defined as the round trip latency for a packet of data. This is an extremely subjective term, and will only be used in broad stroke discussions.

3. **Link Quality** - the potential for lost packets. This is also used to indicate "jitter", or the maximum delay between two consecutive packets being received.

4. **Cost** - the cost per bit to transmit data.

Therefore, the ideal network will provide high bandwidth, low delays, high link quality, and low cost. Currently the existing TDM phone network provides the first three exceptionally well. ATM networks should address the cost issue, and still provide the needed ingredients of high bandwidth, low delays, and acceptable link quality.

	Bandwidth	**Delay**	**Quality**	**Cost**
TDM/PSTN	high	low	high	high
ATM	high	low	med	low
Frame Relay	med	med	high	med
Cellular/wireless	low	med	low	high
X.25	med	med	med	low

Types of Session based fax

There are a few basic features of the different types of transporting fax over data connections. The solution either supports spoofing the T.30 protocol, or it does not. The solution either provides compression of the TIFF data, or not. The solution can run on the host or it must run on the board (primarily due to the number of packets per second).

However, all of the following solutions have a "range" over which they are effective. These ranges vary with the characteristics of the network. The goal of Dialogic's Internet fax strategy is to provide a set of products which adapt to the network in use.

Fax Relay

Fax relay is the protocol which is most closely associated with a Voice Codec. There must be a reliable network with excellent throughput. There is no "fallback" to store and forward fax, and there is no spoofing of the fax transaction. Fax relay merely puts the actual end to end fax protocol on a packet network. Therefore, all pages are confirmed in real time, or the transaction fails. This is "real" realtime.

Because of this, fax relay is a protocol which must run on the card directly. It is the most basic of all the methods described. The fax data, as it is received, is demodulated and placed into packets at a rate of approximately every 20 msec. There is only minor tracking of the actual T.30 protocol to determine when the call will switch between V.17 and V.21. There is no compression of the TIFF data, and little understanding of the HDLC message frames.

NSF transactions are not supported in case a modem scheme is not supported (i.e., not V.27, V.29 or V.17). Otherwise, it would be possible to support.

Because of the lack of image processing, or image knowledge of the fax relay component, the overhead of providing fax relay is roughly equivalent to a V.17 modem algorithm.

Fax Relay will most likely be used in conjunction with voice coders to provide a telco grade Internet telephony edge device. Both voice coders and fax relay have tight timing requirements, and low overhead.

X.38/X.39/FaxPAD

X.38 and X.38 are ITU standards for establishing a connection to send fax traffic over an X.25 network. The original X.38/39 specifications were written in 1993 and revised in 1996. FPAD is the abbreviation for a device called a Facsimile Packet Assembly/Disassembly facility, which is basically what X.38 and X.39 define.

The concept behind the FPAD is to provide a standardized interface between two servers connected to a packet data network. The two servers exchange a standardized set of messages to maintain the connection between to fax machines.

However, there are two very interesting paragraphs in the X.38 specification:

> "Recognizing that the insertion of a FPAD on the PSTN may result in limitations on the interchange of facsimile images between Group 3 facsimile terminals, the introduction of the FPAD should only occur as the result of an explicit act by either the call originator or the call recipient."

Basically stating that the introduction of X.38 technology will introduce limitations between the two fax machines. And second:

> "G3 Facsimile equipment/DCE interface for G3 facsimile equipment accessing the Facsimile Packet Assembly/ Disassembly facility (FPAD)in a public data network situated in the same country."

FPADs have been around for many years. However, there have been only limited deployments until recently. There are one or two companies which are actually providing service based on X.38. However, they are still proprietary implementations of this public protocol.

The significant issue behind FPADs is that spoofing the T.30 protocol is required. While the concept behind spoofing is simple, to send a message to keep the fax device on line while waiting for data on the network, a good implementation takes many years of trial and error.

The trouble is, spoofing uses the part of the T.30 protocol where something went wrong. The well defined section is how to send a fax when everything is right, however, there are very loose interpretations of how a fax machine should behave when things go wrong. Therefore, understanding the characteristics of different fax devices outside the well defined boundary conditions is a tedious process.

Dynamic Realtime

Dynamic Realtime fax is a truly extensible protocol for delivering fax traffic over packet data networks.

	Overhead	Effective Latency	Frame Delays	Issues
Fax Relay	none	10-300 msec	15-25 msec	very low MIPS reqd
FPAD	some	250-1000 msec	250-750 msec	Spoofing causes protocol failures No fallback to store & forward
Dynamic Realtime	some	100-1500 msec	250-1000 msec	large latencies or packet sizes cause spoofed page conf

Hardware Platform

The DM3 hardware architecture is made up of two key products, the high density PCI product and the lower density Venus platform.

The high density platform is a PCI base board with two different daughter board components. The base board has an i960 control processor and variable amounts of global DRAM (2-16 megabytes). The base board also contains four RJ-45 8 wire connections to interface different digital trunks.

The are a series of DSP daughter cards which connect to the base board. The base board must have at least one DSP daughtercard (known as a "hex", for it's six DSPs). There are different daughter cards with different DSP configurations. However, the standard product will ship initially with either a Motorola 603 or 602 DSP running at 66mHz or 100 mHz. It is possible to put up to three daughter cards on a single base board (note, one hex on the base board is a single PCI slot, two or more hex cards will take two PCI slots).

The second daughter card is for the 10BaseT connection. This board will contain a PowerPC chip running an IP stack. This allows PCM data to be received via the TDM network (T1 or E1), processed on the DSPs, packetized and sent out the data network without going to the host. This is a very powerful feature, and should be available by mid 1998.

For more information, visit: www.dialogic.com

Chapter 20

Internet Fax: Optus Software Strategy

As of this writing, most observers are focusing on the Internet as a transmission media. However, the broad availability of Internet fax services makes this aspect of fax implementation a relatively straightforward issue that has little to do with an organization's internal fax services architecture. Any ostensible relationship between enterprise fax software and Internet fax services is more a matter of marketing expediency than it is a technical value-add. The synergy between fax and the Internet that is contingent upon the enterprise fax architecture is browser-based fax server access. What this does is give users the anytime/anywhere access to their faxes. In the case of Optus's LANFax server software FACSys, FACSys Web Agent enables users to point their browsers at a secure Web page, where they can view their faxes, download, forward, etc. This is very convenient for mobile users, and is obviously a better way of communicating than actually dialing into the network—especially if the user is overseas, where the cost of dialing into a modem pool back in the States can be prohibitive. Established fax services vendors such as MCI and Xpedite can also provide value-adds that are currently unavailable in first-generation Internet-based fax services, such as broadcast support.

Server-based rendering

This is in. Whether an organization is already implementing thin clients or has only vague plans to try them in the future, it's smart to employ a fax architecture that executes all rendering at the server. Even if network clients are "fat" PCs, this is still a wise move, since image-rendering tasks consume client CPU cycles. This can become more than just a briefly annoying delay when the documents are large or complex. Server-side rendering also boosts the overall scalability of fax services, since horsepower can be added at a central point to service distributed requests. Integration with thin client solutions such as Citrix WinFrame or Tektronix WinDD is also of great value for organizations attempting to implement more cost-efficient network application architectures.

Ease of administration

One of the most strained resources in any IT environment is network administration. It is therefore critical that all aspects of fax administration—including user rights, queue management and disk space monitoring—be as streamlined as possible. User directories, for example, should be fully synchronized with email administration, so that redundant housekeeping tasks are eliminated. In the Windows NT environment, it is of particular importance that the core enterprise fax application operate as a native NT service, so that operational parameters can be addressed efficiently across a distributed environment. Integration with cost-capture tools, such as those from TRT Technologies is also an important component of administering use of the fax server with the same accountability as the PBX provides for corporate voice communications.

The ongoing importance of fax as an inter-enterprise and enterprise-to-consumer medium is something today's corporate technologists should be careful not to overlook. Many of us inside the industry are so accustomed to email and communicating with people who have the latest upgrade to every software package, that we can lose sight of how the majority of businesses users use technology—especially once you get outside of the United States. Fax is like the mainframe. Client/server and Web-based computing didn't kill it. On the contrary, they have given us new ways to leverage what's already there. The same is true of fax. The key is to implement a solution that allows across-the-board leveraging of existing IT assets while offering the most cost-effective implementation possible—in terms of purchase price, cost-of-ownership and transmission cost reductions.

For more information, visit: www.facsys.com

Chapter 21

The RightFAX Software Strategy

Internet Faxing Philosophy

The Internet has contributed greatly to the unification of business communications. As users try to consolidate their messages, messaging systems race to keep their software competitive. While the Internet is still not as common as the public switched telephone network, businesses are becoming more convinced of its worth and investing heavily in connective infrastructure.

The explosive growth of the Internet's user base has brought Internet technologies to the forefront of the software industry. The computer-based fax industry recognized the Internet to be an opportunity to ensure continued reliance on fax messaging.

Over the past ten years, RightFAX, Inc. has developed its fax server software around the philosophy that the best tools are those that provide choices and control. RightFAX continued to adhere to this logic during the development of its Internet strategy. The result is a comprehensive approach that lets users and administrators choose the way they wish to leverage the power of the Internet in order to unify their communications.

Internet Faxing Benefits

The Internet is associated with many different technologies. For some, the Internet means email or File Transfer Protocol programs. Others know the Internet through the World Wide Web. In fact, the Internet is all of these and more. As the backbone for communications, the Internet is covered by layers of additional functionality and accessibility in the form of email, FTP and the Web. Faxing over the Internet is the implementation of several of these layers.

There is no question that fax messaging is an integral part of business communications. Faxing offers many features that are not offered by email or voice mail. The business world is still document-centric, relying on printers, paper, scanners and copiers for formal communications. Until email and voice mail can include signatures and formatting options, they will never replace faxing.

In addition to the functionality and accessibility benefits, physically transporting information across the Internet instead of the phone lines results in a significant cost savings. The question is how can fax server manufacturers like RightFAX, incorporate the benefits of other messaging systems and use the Internet to enhance fax messaging capabilities.

RightFax 4-part strategy

Using the open protocols of the Internet, including the Web, email, and Internet telephony, RightFAX has developed and implemented a four-part Internet fax development strategy:

1. RightFAX Web Client
2. SMTP/POP3 Internet Mail Gateway
3. Least Cost Routing over the Internet
4. Enhanced Fax Service Providers (EFSP) Connectivity

The philosophy and design of each part is described in detail below.

RightFAX Web Client

One advantage of the World Wide Web is its ability to connect users anywhere in the world and allow them to communicate seamlessly, regardless of the type of machine they are using. Where else can Macintosh, Unix, and Windows machines operate equally as well, in harmony? What other medium allows users to access information from Hong Kong, the UK or right next door with such ease and speed? Businesses wanted this added flexibility and compatibility from their network faxing applications, yet they still needed their LAN clients. Customers did not want to replace the LAN fax system they had invested in, so why not integrate it with the latest technology to extends it's life? The World Wide Web was the obvious answer to the faxing world's demands.

In 1996, RightFAX released the fax industry's first Web Client, designed entirely with the open standards of HTML. RightFAX engineers purposefully designed the Web Client so that it did not require optional plug-ins or scripting languages that were not supported on all of the most popular Web browsers. Regardless of workstation or Web browser type, every user experiences the same functionality and features.

The RightFAX Web Client continues to be a huge success, providing cross-platform compatibility and remote access to anyone with a Web browser. In addition to these benefits, the RightFAX Web Client offers users the convenience of attaching documents in their native formats, including: Word, Excel, JPEG, GIF, and HTML. Administrators enjoy the option to ensure privacy by running the Web Client over a Secure Socket Layer on the Web server and the centralized administrative utilities.

In short, users love it because its easy and convenient to use. Administrators love it because there is no client software to install, minimal setup on the Web server, and easy to manage from anywhere on the network.

SMTP/POP3 Internet Mail Gateway

SMTP/POP3 Internet mail has become the de facto standard for email systems. The most popular proprietary email software have all added SMTP/POP3 connectivity (Microsoft Exchange 5.0, Lotus Notes 4.5, Lotus cc:Mail version 8, HP OpenMail, etc.). RightFAX leverages Internet technology to extend faxing benefits to Internet mail clients, allowing them to create a universal inbox, accessible from anywhere in the world.

With the introduction of the RightFAX SMTP/POP3 Email Gateway, users who were not connected via the LAN or WAN, can now use Internet mail to send and receive faxes, saving long distance phone line charges. Because the gateway relies on the user's email client, there is no additional software to install on each desktop and no training needed.

The SMTP/POP3 gateway can be used in addition to traditional RightFAX clients, so that each user can determine the mode in which they wish to send and receive faxes. The gateway is able to route faxes to email addresses through a variety of methods, including: DID, DNIS, DMTF, etc. Viewing, routing, forwarding, and printing faxes is available at the click of a button. Fax messages can be manipulated through the email client, providing universal inbox functionality.

RightFAX also allows ultimate file format flexibility when sending and receiving faxes. Users can attach files to email messages in their native formats (e.g. MS Word, Excel, WordPerfect, etc.) to be sent as fax images. In addition to receiving fax images in TIFF-G3 or the even better compressed TIFF-G4 encoding, users can receive text versions of their faxes using the Optical Character Recognition (OCR) Module. PostScript and PDF formats are also supported with the optional RightFAX PScript conversion engine. Messages can contain RightFAX embedded codes to add such features as library documents, signatures, or forms, increasing fax automation.

Least Cost Routing over the Internet

The Internet justified least cost routing (LCR) technology for the fax industry. Businesses that were previously limited by their own WANs or LANs could now take advantage of the public Internet to bypass long distance phone lines. Saving money on intrastate, interstate and international calls became a reality for anyone with an Internet connection.

RightFAX extends the functionality of basic LCR systems, offering both sent and received fax routing in it's Intelligent LCR technology. In addition to providing significant cost savings, Intelligent LCR allows RightFAX servers to balance workloads via the Internet. Because RightFAX servers communicate with encrypted packets, all fax traffic is private. RightFAX servers are also able to detect Internet communication errors immediately and dynamically readjust the fax routing appropriately.

EFSP Connectivity

Intelligent LCR offers the unique ability to route faxes to external fax service bureaus, also known as Enhanced Fax Service Providers (EFSP). Business such as FaxSav and Atlas Telecom offer high throughput, guarantees of delivery, high quality, and reduced costs as compared to the public switched telephone network by eliminating most long distance charges. An EFSP can usually deliver thousands of faxes in a matter of minutes or hours because they have hundreds or thousands of fax ports in concurrent operation.

Additionally, EFSPs have developed a large, global network. In many cases, their infrastructure extends to thousands of sites. Many companies rely on their services because it is not possible or feasible to have an internally owned and operated server in every country, state, or city to which they send faxes. EFSPs are able to offer a reasonable alternative to such businesses.

EFSPs also offer a variety of messaging services international faxing, fax broadcasts, fax-on-demand and mailbox delivery through their Internet connection. Because faxes are routed over the Internet, every fax begins as a local call.

Making It Work for You

RightFax Internet faxing strategy lets businesses decide how best to leverage the power of the Internet. By adhering to Web and email open standards, RightFAX makes it easy for users to choose the method of communication they prefer. By incorporating Internet into least cost routing technology, administrators have the flexibility to route received and sent faxes via the most economical way, without inconveniencing users with awkward dialing patterns or routing strings.

The Internet is not a threat to traditional fax servers, it's an opportunity for enhancement. And RightFAX is poised to take advantage of that opportunity.

For more information, visit www.rightfax.com

Chapter 22

Open Port Technology's ISP Internet Fax Strategy

Introduction

Internet Service Providers (ISPs) face the reality that in a highly competitive market, increasing the return on their network assets is the best way to sustain growth and achieve profitability. Market analysts agree that ISPs must offer value-added services to subscribers to remain competitive and limit churn. To produce optimal revenue results, new applications should:

- Provide a method of driving revenue-generating transactions;
- Leverage existing network resources;
- Differentiate ISPs from competitors;
- Add a new transaction-based billing paradigm;
- Optimize bandwidth usage by end-users;

- Augment—not compromise—the quality of applications and services;
- Minimize incremental investment.

Fax messaging is the logical first step to establishing value-added service offerings that meet these criteria. Fax as a revenue generating service is ready for the Internet now and moves easily into an ISP's environment. Fax over the Internet has been proven to reduce faxing costs for end users while adding revenue to the ISP.

The Harmony NSP fax solution from Open Port Technology provides the scalability and reliability that ISPs want, while focusing on the future of messaging: the universal mailbox that contains all messaging types, such as voice, fax, email, and video messages. Harmony NSP offers an end-to-end Internet fax solution, including both service providers and end users. Open Port not only provides an application platform, but also supplies the means to get transactions from the end-user to the service provider with Customer Premise Equipment (CPE) and software.

Corporate Demand for Internet Fax

Internet and IP-based Intranets and Extranets are fast becoming an important backbone component of corporations' messaging infrastructures. The telephone network and data networks (e.g., private wide area networks—WANs—and shared virtual private network—VPNs) that have existed side-by-side for decades are converging into a single communication network that offers powerful and economical new communications options. Fax is essentially data which has traditionally been carried by telephone networks, making fax a prime candidate for initial application integration. Corporations look to ISPs to lead the convergence of computers and telephony with value-added services.

Business Case

The Backbone Provider derives fax traffic from two sources: their own end-users and the Service Providers that re-sell the fax services to their end-users.

Backbone Providers deploy Harmony NSP globally on their network infrastructure to take advantage of their world-wide POPs. The Harmony NSP solution enables the Backbone Provider to build a profitable fax business by marketing Internet fax services to end-users in fax intensive industries. Equally important, they resell to their Service Providers, who need international delivery points. The end-users and Service Providers funnel fax traffic up to them for routing, transport and final delivery.

Open Port offers a separate solution to the Service Provider through the reselling of different Harmony NSP components by the upstream Backbone Provider, or by VARs. Fax-enabled ISPs offer the benefits of Internet faxing to their end-users who can extend their network faxing capabilities far beyond their own LANs and WANs by deploying the Harmony Enterprise system with an Internet gateway.

Harmony NSP Architecture

Backbone Providers require that fax transactions move quickly and easily around their networks. The two main components of the Harmony NSP system, the Hub Server and the Fax Access Server (FAS), separate fax processing into three operations—routing, transport and final delivery—for unequaled response times.

A backbone Hub has the Hub Server handling the routing and transport of fax traffic, and sending messages for final delivery to the correct POP, contains the Fax Access Server (FAS) and third party Remote Access Server (RAS).

The Hub Server handles the administration of fax traffic, as well as the routing and transport of faxes as needed by the ISP's network architecture. For the latter, the Hub Server looks at an incoming fax address, and routes it to either a mailbox for pick up by a user, or re-directs routing and transport to a local Fax Access Server, for delivery out to the Public Switched Telephone Network (PSTN). The administration and transaction logging functions of the Hub Server can be physically placed in an ISP's Network Operations Centers (NOCs), for ease of system control. Maintenance for the entire fax-enabled system can then be performed in the same manner as other maintenance functions performed by NOC personnel.

The FAS, along with the third party Remote Access Server (RAS) handles delivery to the PSTN of off-net traffic. The FAS/RAS pair enables the ISP to leverage the RAS as an existing network resource. A fax-enabled POP must have a FAS/RAS "pair" (Note 1: One FAS can address ports on multiple RASs; this paper refers to "pairs" only in that the RAS needs a FAS to drive the fax communication with a fax machine). End-users can dial in to the RAS to submit faxes. If the end-user is at a fax machine, the FAS performs the necessary protocol translation of fax to data (Note 2: The ISP can, if it does not want to re-use its existing RAS equipment and phone lines, substitute Open Port's External Communication Devices (ECDs), which use standard telephony and fax boards supplied by Dialogic and Brooktrout, in place of the FAS/RAS pair.). If instead, the end-user is at a desktop computer, the communication is forwarded directly to a Hub Server.

Service Providers also participate in the fax services offered by the Backbone Provider. The addition of re-seller traffic gives the Backbone Provider enhanced revenue for dollars invested in the Harmony ISP system. Similarly, corporations deploy the Harmony Fax Server with an Internet gateway to link to an NSP's Hub Server.

Scalability Through Modularity

The Hub Server and FAS contain the software modules that comprise the main Harmony NSP system. Because Harmony

NSP is a fully distributed system, it provides unprecedented scalability through the flexibility of the modules. "Fully distributed" means that the modules are logically connected, not tied to any specific location. The module functions may be distributed across multiple servers, so more than one server could be dedicated to running a single process. While the main functionality of the servers is fixed, certain modules can be installed at the Hub or POP level, giving the Backbone Provider the ability to distribute certain functionality of the system as it chooses.

Another level of scalability comes from the ability of the FAS/RAS pairs at a POP to support inbound/outbound ports, and each Hub Server to support several FAS/RAS pairs. The FAS/RAS pairs can be in the same or numerous remote physical locations, depending on the needs of the ISP. It allows processing to be distributed across the network for huge through-put potential.

Manageability

Harmony NSP allows ISPs to determine the best way to administer the fax system. It supports a centralized topology "star" configuration with a large central processing center. All (or most) of the Hub Servers can be located at the central administration site, with FAS/RAS pairs anywhere the ISP needs them, in any combination.

Harmony NSP also supports the alternate approach of decentralized topology, with multiple administration sites handling the same number of POPs as the centralized model. In both topologies, the FAS/RAS pair deliver faxes from any Hub Server or other FAS/RAS pair, regardless of where a fax enters the system.

Fax Traffic Flows

Harmony NSP handles the three main traffic flows:

1. General flows – the typical, daily fax traffic, through the Backbone Provider and Service Provider systems, that

users send and receive. The end-user systems give the ability to send faxes by fax machine, desktop client, or email package. Users have the ability to receive both Internet and traditional faxes via fax application mailboxes (Open Port CPE), at fax machines, or in a properly configured email box.

2. Outbound broadcast flows -ability of an ISP to handle large volumes of fax traffic at once. Harmony NSP has the option of creating cover sheets at the delivering POP, saving tremendous amounts of bandwidth. Without it, a 10,000-recipient, three page fax becomes a 10,000-recipient, four page fax going from through the network.

3. High volume inbound flows - At the end-user level, companies may require an application for mailboxes to handle significant amounts of inbound fax traffic. With the mailboxes located at the POP, they access faxes locally, or have the option to store them on their own Harmony server.

Interoperability

Open Port is committed to an open architecture with its software/hardware solutions. Interoperability is key in the ISP market since potential standards are still being discussed. Open Port is committed to hardware independence. Open Port solutions remain flexible to fit in an ISPs existing network infrastructure.

Interoperability extends to the end-user level as well. The Harmony ISP solution supports popular operating platforms used by corporations, and an ISP's customers are able to connect to its Internet fax service with their currently deployed hardware and software.

Enabling Service Providers: Service Provider Architecture

The Service Provider system, while retaining the transport features of the Backbone system, provides for the direct connection of Open Port's standalone fax clients, browser client, and the user's own email client for sending and receiving faxes. Faxes sent to and generated from standalone fax machines are also included in the system.

The Service Provider has the option of having their Harmony NSP-enabled Backbone Provider perform all or a portion of fax transaction deliveries, allowing the Service Provider to outsource as much of the delivery mechanisms as they wish.

Gateway Server

Service Providers deploy the Harmony NSP Gateway Server, which allows the direct connected business end-users to connect LANs to send and receive faxes from desktops and connected or standalone fax machines. Additionally, standalone PC clients and fax machines are able to participate, serving the SOHO and consumer markets. The Service Provider system, like the Backbone system, has no limits on the types of end-user traffic.

The Gateway Server is the conduit for fax traffic between the Service Provider's end-users and the Service Provider's Backbone Provider. End-users may or may not have an Open Port Fax Server on premise.

- Standalone Clients: When the Gateway Server is handling standalone client users, authentication and storage functions, such as mailboxes, cover sheets, and shared phone books are handled by the Gateway Server. User administrative functions such as adding, changing and deleting users, are performed by the Service Provider on behalf of the end-user.
- End-Users with LAN-based Open Port Fax Server: When the end-user has its own Open Port Fax Server attached to its LAN, the Fax Server performs the storage and user administration functions locally, but the traffic still passes through the Gateway Server. (Note: A Service Provider with no means of offering direct connections

to multiple customer premises-based Fax Servers can implement the Gateway Server without this component.)

Users of both types are authenticated by the Gateway Server at the point of log-in. After log-in and fax initiation, a fax flows up from any client through the Gateway Server to delivery to either:

1. Any Gateway Server on the Service Provider's network (for desktop-to-desktop delivery);
2. FAS/RAS "pair" (defined below) for off-net delivery; or
3. The Backbone Provider's cloud for further routing and final delivery.

Harmony NSP for the Backbone Provider: Full System Features and Compatibility

Billing

Harmony NSP allows the ISP to use its own billing system to charge customers for fax usage. An additional module, the Log Server Module, includes detailed tracking capabilities through call detail records (CDRs) for ISP billing purposes. CDRs can be collected from delivering servers, or kept in a central location, at the ISP's discretion. The billing files are in ASCII format for easy conversion into any administering system. Items tracked include fax duration in bytes, minutes and pages. The customer information tracked is extensive.

User Authentication and System Monitoring

ISPs need to be sure that users are legitimate. The Harmony NSP Login module performs authentication for any application logging into the system. This system authorizes users accessing the network, ensuring that incoming faxes are valid. Harmony NSP allows system monitoring by generating SNMP (Simple Network Monitoring Protocol) "traps". The system, depending on the level of service desired by the Provider, is supported for details on system status, such as fault identification and statistical information.

Receive-Side Capabilities

A complete fax system requires both send and receive functions. Harmony NSP offers the ISP full inbound capabilities. Receive-side benefits extend to both ISPs that would otherwise lose important potential revenue, and to mailbox owners. Mailbox revenue can be generated either from mailbox rental, per transaction, or both. Mailbox owners benefit from the privacy a mailbox provides, and worldwide access to their faxes.

Harmony NSP can be scaled to ensure "never busy." With enough ports for the customer base, the mailbox owner's fax number is never busy, minimizing sender and receiver frustration at busy fax lines. For an ISP that is also a telco, "never busy" fax numbers result in more call completions, reducing operating expenses.

Notification

End-users want to know the status of sent and received faxes, answering the questions of "Did my fax go through?" and "Has my fax come in?" Harmony NSP notifies the user of sent and received faxes in several ways. Fax machine senders are notified of success or failure by receipt, which contains a reduced copy of the first page of the fax. The desktop client allows users to check the status of sent faxes, and are informed of inbound faxes or outbound receipt by either pop-up window or email. Desktop fax senders who use email software to submit faxes are notified via email.

File Conversion

Harmony NSP was designed to optimize an IP network. ISPs are able to offer value-added fax service to customers with minimal impact on their internal network performance, while increasing the bandwidth requirements of their end-users through the increased traffic fax brings, thus escalating ISP revenues.

The desktop client user prints to the Harmony fax print driver, which creates a Postscript output file—at about half the size of the equivalent "TIFF" image file that represents the fax—and submits it to the Harmony NSP Hub Server. The Server routes it according to the least cost path (and note that the least cost routing mechanism allows the ISP to consider network costs as well as phone costs). With this feature even larger broadcast fax transactions flow easily through the system since client originated faxes are not converted to TIFF format until they reach the delivering FAS/RAS pair. Also, with its modular design, Harmony NSP balances the flows across servers during periods of heavy use whether inbound, outbound or a combination of both. So, network bandwidth is used when it is available, with peak loads stretched out as the network and service level guarantees will allow.

Having less internal network bandwidth consumed by each transaction enables the ISP to drive more traffic across the network without requiring more resources, thus increasing return on investment. End-users with their increased traffic traveling over their limited access dedicated lines will require additional bandwidth. While they may require additional bandwidth, they realize overall cost savings due to reductions in fax bills for traditional PSTN-based fax.

Sockets Communication

Sockets-based communication gives the Backbone Provider (and therefore end-user) greater confidence that the system will not lose faxes due to collisions of data over the network. Harmony NSP utilizes sockets-based sessions on both the LAN and the Internet levels. "Sockets" communication in Harmony NSP occurs when a fax message is sent from one processor to another, the receiving server confirms to the sending server that it may send the fax. The sending server releases the fax only after receiving the proper commitment. The connection is dropped after the fax is complete.

Note that this compares favorably to email based communications, where the originating server simply sends the fax without getting commitment from the receiving server that it takes responsibility for delivery. The sending server may—or may not—know the fax was eventually delivered, often requiring a polling solution that can traverse multiple network hops. Harmony NSP's sockets approach eliminates the problems of multiple hop deliveries associated with e-mail protocols, as well as the needless polling—using unnecessary bandwidth—to update message status records.

Transaction-Based Network Service

Internet fax, as enabled by Harmony NSP, offers an enhanced network service that is billed based on actual customer usage measured in bytes, minutes or pages, rather than the high risk of fixed monthly rate Internet services for an anticipated average user that may be highly underestimated. A per transaction Internet-based fax service can provide an ISP extremely high margins on both domestic and international fax traffic.

Open Port's Internet Fax Vision

Open Port's Harmony NSP strategy is based on a vision of the future of Internet fax messaging being dominated by the Backbone Providers. Other Internet fax approaches see ISPs playing a role similar to what is found with Internet email. In the email world, ISPs of all shapes and sizes exchange traffic, with little difficulty. However, Open Port believes strongly that fax is quite different than email.

- Email messages never go "off-net". There is no phone call cost for email delivery. Fax messaging requires controls that can recover these costs.
- POP requirements for email messaging are no different

from those for "generic" Internet usage. Fax delivery requires a configuration at POP sites that can handle fax traffic as distinguished from data traffic. The investment to add email as a service offering is less than the investment to add fax. This being true, the ISP is not satisfied unless there is a mechanism to pay for its investment costs.

- The Backbone Providers are significantly better positioned to offer a level of service that has come to be expected among fax users—virtually instant delivery—that has not been the expectation of email users. Backbone Providers can guarantee delivery because they can be sure that the message will only travel on their own networks, while independent ISPs, relying on email based standards, will be at the mercy of the Internet email delivery system.

Open Port believes that the Backbone Providers will become synonymous with quality Internet fax transmissions, that there are multiple Backbone Providers capable of doing so, and their level of service will be so superior to independent local Service Providers that users will have no difficulty choosing the Backbone Providers. The optimal position for the local Service Provider is to accept one (or more) of the Backbone Providers as its "fax transport service", just as they accept them today as their data transport service. The larger of these Service Providers may wish to deploy servers within their own geographic areas.

Conclusion

The Internet messaging environment is reaching a critical stage. Because fax and other messaging types represent a giant revenue opportunity that is currently realized almost exclusively by traditional telephone service providers, ISPs must choose which messaging solutions will take them into the next century. Fax messaging, with Harmony NSP, generates revenue now, while setting up an ISP for the burgeoning universal messaging market to come.

Leading ISPs will build messaging infrastructure—own and control it—and demand stability and versatility with their chosen system. Harmony is a proven solution into which companies worldwide have successfully placed their mission-critical traffic. Harmony NSP offers the best solution, well suited for the Backbone Provider.

For more information, visit: www.openport.com

Chapter 23

Panasonic's Internet Fax Machine

The PanaFax UF-770i sends messages via the Internet. With it, you can send faxes directly to any email account, computer with an Internet connection or other Internet Fax Machine. The Internet Fax Machine also has G3 fax compatibility, as well as network scanning capabilities, allowing it to easily integrate within existing network/email environments.

Sending a fax via the Internet with the UF-770i is no different than sending a fax from an ordinary fax machine. Just enter the email address and press Start. All necessary data conversion is done automatically, so you don't even need to know whether you're sending a fax or email.

The Internet-Ready Paper-Based Communications Solution

The UF-770 Internet Fax Machine supports a standard 10baseT Ethernet connection (RJ-45) so you can plug it directly into your LAN. It also comes with a telephone jack for communication with regular G3 faxes. By taking advantage of the built-in Internet protocols (TCP/IP, SMTP, MIME) and assigning a unique IP address to the machine, the unit can be used to send/receive information to a conventional G3 fax, an email-capable remote PC, or another Internet Fax Machine. In addition to flexible relay/forwarding features for easy integration into existing email and fax environments. This creates a streamlined, seamless information infrastructure without reengineering your current communications network.

The UF-770 Internet FAX uses your existing email environment to transmit documents. The result is reduced communications costs. Simply plug the UF-770i into your network hub to transform your information network into a much more flexible, easier-to-use and more productive communications system.

Email Transmission, Fax-Style

The UF-770i sends a TIFF file as an email attachment to another Internet Fax Machine or a PC with an email address. Data received by a PC can be displayed with a TIFF viewer, providing paperless fax communications.

Forwards Fax and Email Received in Memory to Any Destination You Specify

The UF-770i can memory-forward received data as email with attached fax data, to a fax or PC. The Internet Fax Machine receives the image data from G3 faxes and email via the LAN first, then forwards the data directly to a preregistered fax or PC.

Increased Integration with Relay/Forwarding Features

The UF-770i's Relay/Forwarding capabilities can even be used as a LANFax server and a G3 Fax Gateway.

Network Scanner Capability

By pre-registering addresses of computer terminals on the One-Touch Keys, you can scan images as TIFF files and send them straight to your PC. Unlike conventional scanners, the scanning procedure is quick and simple and the transport to your PC is done via email—just press a key on the Internet Fax Machine. There's no need for back and forth communication between the fax and your PC.

Specifications

Compatibility:	ITU-T Group 3, ECM
Document Size:	Max. 11" x 78.7" (280 mm x 2000 mm) Min. 5.8" x 5.0" (148 mm x 128 mm)
Resolution:	STANDARD: 203 dots/inch x 98 lines/inch FINE: 203 dots/inch x 196 lines/inch SUPER FINE: 203 dots/inch x 391 lines/inch
Printing resolution:	406 dots/inch x 391 lines/inch
Transmission speed:	Approx. 6 seconds page
Recording method:	Laser printing on plain paper
Recording paper size:	Letter/A4/Legal, cut-sheet plain paper
Paper Supply:	250 sheets (optional cassette of 500, 750, or 1000)
Modem speed:	14400/12000/9600/7200/4800/2400 BPS with automatic fallback
Power requirement:	90-138VAC, 47-63 Hz, single phase
Dimensions	(WxDxH): 16.9" x 16.4" x 11.0"
Weight:	33
Recording speed:	10 ppm
Memory Capacity:	up to 40 pages
Optional memory card:	Base memory plus up to 155/235/405/740 pages*4
Received Email width:	A4 size only

Optional Accessories: Telephone handset (UE-403117)
250-sheet Universal Paper Feed Unit (UE-409057)
500-sheet Universal Paper Feed Unit (UE-409056)
1/2/4/8 - MB Memory Cards
72-hours Battery Back-up Option Kit (UE-403125)
PanaFax LAN Interface Kit (UE-404063)

The UF-770I's software, reference manuals and TIFF converter are available online. To download, or for more information, visit: www.panasonic.com

Section VII

Building Internet and Computer Based Fax Applications

Chapter 24

SOHO & WorkGroup Internet Fax

with Brooktrout's IP Fax Router

Brooktrout's IP Fax Products

Store-and-Forward

Brooktrout's store-and-forward products include an embedded CPE system, the IP/FaxRouter, and a Windows NT utility, the NT/FaxRouter.

The IP/FaxRouter, was one of the first IP fax products released. It is a small (approximately the size of a dictionary) embedded system that requires no user configuration beyond the assignment of an IP address. IP/FaxRouters act as "nodes" or "gateways" on an IP fax network. It is connected to a conventional telephone line, and to fax machines or a PBX which allow many fax machines to be supported by a single IP/FaxRouter. It supports sending faxes to any fax connected directly to the IP/FaxRouter or an attached PBX.

It can also deliver "off-net" by completing calls over the PSTN.

An IP/FaxRouter network is configured and managed by a central management utility, CNMS. CNMS is used to assign routing tables to every node on a network. It also collects, tracks and reports activity at each node on the network.

To support "IP fax enabling" computer-based fax applications (such as LAN fax servers and fax broadcast systems), Brooktrout made its router technology accessible to system developers by introducing the IP/FaxRouter API, which allows a fax application to submit faxes to and receive faxes from the IP/FaxRouter.

Later in 1997, they released the NT/FaxRouter utility. This Windows NT utility provides the core fax packetization, routing and management capability of the IP/FaxRouter (see Figure 1). It submits faxes to and receives faxes from a conventional fax application (for example, a fax server based on Brooktrout's TR114 Series™ multichannel fax boards) via a DLL. The NT/FaxRouter server acts as an IP/FaxRouter node and is managed by CNMS. This interoperability facilitates scaling of IP/FaxRouter systems and mixing of service provider and CPE systems.

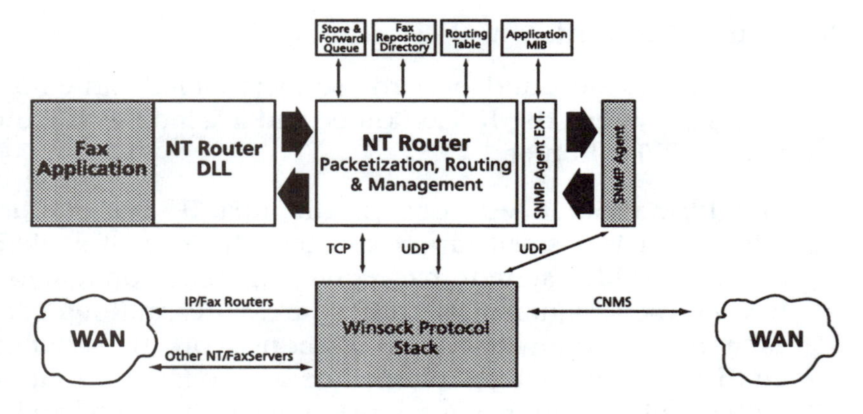

Figure 1. NT/FaxRouter SDK with CNMS Compatibility

Real-Time

Brooktrout has been active in real-time transmission of facsimile over packet data networks since 1991. Customers such as Graphnet, Inc., Telestra Communications Ltd., SoftLinx, Inc., and Voice and Data Systems have all developed real-time services or systems based on Brooktrout's FaxPAD. FaxPAD, derived from ITU standards X.38 and X.39, was originally developed for use on X.25 networks. Recently, FaxPAD has been modified for use on IP networks.

In late 1997, Brooktrout released new tools to simplify the development of FaxPAD systems. The FaxPAD SDK is an extension of Brooktrout's fax and voice application programming interface (API) that incorporates a host-based state machine to provide call control and fax protocol synchronization of the gateway servers. Brooktrout's TR114 Series multichannel fax and voice boards incorporate modifications to the T.30 fax protocol which support the "spoofing" that enables FaxPAD to provide reliable service over networks with longer delays.

Also in late 1997, Brooktrout released software and hardware for FaxRelay. It will be provided as an option to Brooktrout's IP voice products, SpeechPak software and the TR2000 Series™ DSP resource board.

These IP/Fax products are part of a boarder Brooktrout product line that addresses IP/Telephony.

For more information, visit: www.brooktrout.com

Chapter 27

SOHO LANFax

with Castelle FaxPress Fax Server

The FaxPress family of products offers a flexible and scaleable approach to integrating fax capabilities into workgroups and departments of all sizes. FaxPress is a complete software and hardware solution that plugs directly into the network and attaches to multiple fax lines, bringing comprehensive fax services and sophisticated fax management features to local area network users. All products are fully self-contained solutions with all the necessary hardware and software. They do not require dedicated PC's, 3rd party network or modem boards. Additional servers products can be added to the network at any time to handle increased fax demand automatically. Network administrators can provide network users with "Universal Inboxes" where all incoming faxes, email and other communications can be centralized and easily managed. With optional interfaces to the most popular email packages such as Microsoft Exchange, Lotus cc:Mail, Lotus Notes, Novell GroupWise, and SMTP-compatible mail systems, FaxPress allows users to send and receive faxes directly through their email systems.

FaxPress delivers faxing capabilities to either Microsoft Windows NT or Novell NetWare environments and can be deployed in heterogeneous (mixed) networks where both TCP/IP and IPX/SPX protocols are present (optional). Fax Services can be accessed from any DOS, Windows, Windows 95, and Windows NT workstation on the network. An Internet option allows access to fax services using a standard Netscape browser and a network connection (local or via Internet).

FaxPress Server Overview

The system's hardware combines an intelligent network interface, a raster image processor, a system CPU and a number of fax modems. Every server is delivered with Castelle's Client/Server Software, which allows the server to integrate with leading network environments such as Microsoft Windows NT and Novell NetWare.

All fax servers support the T.30 sub-addressing standard to automatically route incoming faxes to individual workstations. Incoming faxes may also be delivered to the end-user's workstation via DID (direct inward dialing), which funnels a range of incoming fax numbers over a single DID trunk line, or through line routing. Each modem can be configured for incoming only, outgoing only, or bi-directional operation.

Castelle fax servers are scaleable systems that have been designed for growth and company-wide utilization. As fax demand grows, additional fax servers can be added to the network without having to replace existing investments. Automatic load balancing simplifies management and administration of multiple fax servers and allows the increased load to be distributed automatically across multiple fax servers. In high-demand fax environments, up to five models can be deployed concurrently, supporting up to 20 simultaneous phone lines.

FaxPress 1500, 1500-N and OfficeConnect Fax Server

The OfficeConnect Fax Server is an integrated component of the 3Com/Castelle OfficeConnect family of products designed specifically for the small business and branch office environments (under 50 users) with low to medium fax traffic. The OfficeConnect fax server comes equipped with 4MB of on-board memory and a single high-speed 14.4Kbps fax modem. It can be connected to either a Novell NetWare or a Windows NT server to deliver fax services to a local workgroup.

The FaxPress 1500-N is designed for small to medium-size Novell or Windows NT networks and can accommodate heavier fax traffic. It comes equipped with 8MB of on board memory and is available with up to two fax lines per unit. The FaxPress 1500 is a lower cost unit that comes equipped with only 4MB of memory and is designed specifically to connect to a Novell NetWare server. The FaxPress 1500 can be optionally upgraded to 8MB of memory and support Windows NT and NetWare simultaneously.

High-Performance FaxPress 3500 and 3000 Servers

The FaxPress 3500 and 3000, the most powerful and feature-rich members of the family, are available for either Ethernet or Token Ring networks and can support up to four separate fax lines per unit. They use 2 (or 3) Motorola 68030 microprocessors and are configured with high-speed Rockwell International 14.4Kbps fax modems. The 3500 is equipped with 8MB's of memory (upgradeable to 12MB) and designed for high-speed fax rasterization at the server. The 3500 can be connected to either a Novell NetWare or a Windows NT server. The 3000 is a lower cost unit that comes equipped with 4MB of memory and supports Novell NetWare environments only. It can be optionally upgraded to 8MB of memory and support Windows NT and NetWare simultaneously.

The 3500 and 3000 are ideally suited to application environments with abundant faxing needs that can completely utilize all four fax modems. These units allow all four lines to transmit simultaneously. In broadcast applications, 4 to 6 single page faxes can be transmitted in under one minute. There is no need to wait for the documents to be converted from a PCL image to a fax image, since these units come equipped with higher speed processors that can convert documents faster than four modems can transmit.

Network Integration and Utilization

FaxPress is a true network appliance dedicated to faxing—it attaches to the network and provides fax services to all designated users on the network that have been given permission to access its services. FaxPress also relies on the availability of other services on the network. In particular, the availability of a network file server for storage of configuration files, incoming and outgoing fax queues, phone books, user data base and executable files. FaxPress can take advantage of the file services offered in either Novell NetWare or Microsoft Windows NT server environments.

Communications between the client workstations and FaxPress are conducted on a peer-to-peer basis via either the IPX/SPX protocol (NetWare clients), TCP/IP protocol (NT clients), or both (add-on required). The result of this advanced architecture is that users are not required to have login rights on the file server used by FaxPress in order to take advantage of fax services.

Load balancing allows additional servers to be added to the network at any time to handle increased fax demand. This is particularly useful in fax-on-demand and fax-broadcast applications. Any number of servers can be configured in a "Slave" mode, allowing them to share the load of outgoing faxes from a designated "Master" server. Also, different models can be mixed and matched using load balancing, allowing customers to retain their investment in hardware. When not in "Slave" mode, the server configured as an independent fax server with its own incoming and outgoing fax queues.

FaxPress Functionality

FaxPress has been designed for Microsoft Windows NT and Novell NetWare 2.x, 3.x, and 4.x environments. It integrates with NetWare 4.x Network Directory Services (NDS) and eliminates the need for bindery emulation in the directory.

FaxPress can communicate natively with network PC's using the IPX/SPX protocol—which is typically found in NetWare networks, OR using the TCP/IP protocol—which is typically found in Windows NT and enterprise networks. Using the Multi-protocol Software Option, it can be configured to support mixed client environments by communicating simultaneously using both TCP/IP and IPX/SPX.

FaxPress supports various network protocols including:

- 802.3 & 802.2
- SNAP
- 8137 Ethernet II with programmable network type
- Source and non-source routing in Token Ring environments
- Configurable Token Ring broadcast type: all routes and single routes
- Adjustable packet size

Network Traffic Considerations

During normal operation, FaxPress does not generate significant additional network traffic on a typical network. A fax server, however, like other devices operating with images, can increase the volume of network traffic when moving large faxes around the network. FaxPress performs all fax-related image processing internally, drastically reducing the need to move large bit maps between client workstations, file servers, and the fax server. A centralized database further reduces the need to move data over the network.

Fax Printing Options

FaxPress provides significant flexibility in printing and forwarding of incoming faxes. All servers are equipped with a local printer port which can be connected to a printer to automatically print incoming faxes as they are received. The local printer port can also be configured to function as a Network printer. This allows the printer connected to the server to be used to print regular print jobs as well as the incoming faxes. Furthermore, the server can be configured to send incoming faxes to any network printer on a user by user basis. This allows users to receive their faxes on a printer that is within close proximity of their work area. System administrators can select to have incoming faxes printed automatically or forwarded to a designated user for proper forwarding. Users with personal fax numbers can opt to have their incoming faxes printed automatically or stored by the fax server for private viewing.

Security

FaxPress provides several layers of security:

- Network security
- Fax server access security
- PBX dial out password security
- Incoming fax access security and transaction logs and notices access security

Supported File Formats

FaxPress supports several outgoing file formats in which up to 64 files with various formats can be combined in a single fax. In Windows, the files in the valid format can be attached directly to the active document. Formats include: PCLL®5 (all the features of HP LaserJet III and majority of the LaserJet 4 features are supported), ASCII, PCX, CFX (saved faxes), DCX (email) The FaxPress can also support several incoming file formats that can be seen on the viewer. These include: RLE, BMP, CLP, TIF, MSP, FAX.

Management of Outgoing Faxes

FaxPress gives users and system administrators a wide-degree of flexibility and control over outgoing faxes. Users are allowed to perform management operations only on their own faxes. System administrators with proper privileges can perform management operation on all outgoing faxes in the queue. Management operations include:

- Monitor outgoing fax queue, located on the file server, with real time status updates
- Resequence faxes in the outgoing queue
- Put outgoing faxes on-hold
- Removing the hold status on an outgoing fax
- Cancel outgoing faxes
- Change the phone number
- Display failure information for failed outgoing faxes
- View failed faxes
- Resend failed faxes
- Automatically print outgoing faxes or confirmation pages as they are faxed on local or network printer
- Review the list of notices for each user indicating the status of submitted fax jobs

Routing of Incoming Faxes

FaxPress supports several different ways to route incoming faxes to individual workstations and/or printers (beyond the Secure Manual Routing Method):

- DTMF (Dual Tone Multi-Frequency) automatic routing
- Line routing
- DID (Direct Inward Dialing) automatic routing
- T.30 sub-addressing standard

DTMF routing method does not require any special equipment or phone lines, but it is not very convenient for the sender.

Line routing (assigned user routing) - This option allows for the ability to assign an individual user and print queue to each incoming telephone line. All the incoming faxes from that line will be printed on the network printer servicing that assigned print queue. The assigned user can administer the incoming faxes received on that line. FaxPress maintains a list of notices for each user indicating the reception and routing of faxes.

Direct-Inward-Dialing (DID) - this technology was developed by AT&T for use with PBX's. Summarized, it allows multiple phone numbers to be placed on a single phone line. With DID every user gets individual phone numbers. The last 3 or 4 digits of that number correspond to the individual's extension. When the telephone company's central office places a call via the DID line, it will forward these digits to the receiving device.

The FaxPress software will recognize that those digits correspond to a defined FaxPress User and then it will direct the ensuing inbound fax to that user's FaxPress incoming mailbox.

Unlike a regular phone line, the DID trunk expects the receiving end of the call to provide the battery feed that makes the line usable. This requires an interface, such as the EXACOM DID Interface Module, to provide the "link" between standard DID trunks and regular phone line inputs of FaxPress. The EXACOM interface translates DID signals into DTMF signals recognized by FaxPress.

T.30 Sub-Addressing Standard - The subaddressing is the new standard for routing of the incoming faxes. The subaddress is added to the fax number separated by #s, i.e.:

408-492-1964#0252#0223#0268#0370.

FaxPress will recognize the sub-addresses and route the incoming fax to the user with matching ID.

In order for sub-addressing to function, both transmitting and receiving systems must have the capability. Most fax machines do not yet support sub-addressing. Sub-addressing first became available in fax machines during 1995, and is now becoming available between computer based faxed system. This sub-addressing feature can easily work between two FaxPress fax servers. If a fax is sent from a fax device without sub-addressing capabilities it will be received in the unaddressed fax queue.

Management of Incoming Faxes

FaxPress gives users and system administrators a wide-degree of flexibility and control over incoming faxes. Users are allowed to perform management operations only on their own faxes. System administrators with proper privileges can perform management operation on all incoming faxes in the queue. Management operations include:

- Automatic alert/notification upon receipt of an incoming fax
- Choice of storing incoming faxes on the file server and/or printing to a predefined print queue
- Real-time monitoring of the incoming fax queue
- Viewing of incoming faxes on workstation
- Refaxing an incoming fax to other number(s)
- Printing on local or network HP LaserJet compatibles and/or Postscript printer
- Rerouting to other users on the network using Castelle utilities or email
- Deleting received faxes
- Edit incoming faxes
- Text annotate incoming faxes
- Optional OCR software - it will allow the user to covert incoming faxes into editable text
- Save faxes in the following formats:

Castelle Format	*.CFX	MS-Paint	*.MSP
PaintBrush	*.PCX	(multipage PCX)	
Tag Image File Format	*.TIF	Windows Clipboard Bitmap	*.CLP
CCITT Group III FAX	*.FAX	Windows Compressed	*.RLE
CompuServe	*.GIF	Windows Bitmap	*.BMP
OS/2 Bitmap	*.BMP	Windows Meta File	*.WMF

Transaction Logs

FaxPress maintains transaction logs of all incoming and outgoing calls. Every month two files are generated—one for transmitted faxes and one for faxes received. The supervisor can view, print, save, or delete transaction logs.

Email Integration

FaxPress can integrate seamlessly with most of the leading email systems available on the market today. With an email gateway, FaxPress provides email users with a convenient and transparent way to send, receive, and view faxes from within their email application. FaxPress provides status of outgoing transmissions and informs users of incoming faxes via email messages. Received faxes are sent to their recipients as email attachment and can be viewed directly from the email application.

Castelle provides FaxPress gateways for the following email environments:

- MHS compliant systems (NetWare)
- Lotus cc:Mail (NetWare)
- Novell GroupWise (NetWare) (Available in Q3'96)
- Lotus Notes (Windows NT) (Currently in Beta)
- Microsoft Exchange (Windows NT) (Available in Q3'96)

Chapter 26

Workgroup and Enterprise LANFax Faxing

with Omtool's Fax Sr.

Fax Sr. for Intranets is an enterprise network faxing system, designed to fully integrate with a company's Intranet and Internet strategy. Fax Sr. delivers users seamless web-based faxing, while allowing administrators flexibility in Intranet design.

Browser Clients

With the Fax Sr. web clients, users can manage their fax mailbox from any browser. Both the ActiveX and JAVA Fax interfaces deliver secure access from any location or desktop worldwide. Users can send faxes plus file, delete, annotate and view all received faxes. With one easy to use interface, all fax functions including phone-book administration can be performed. Sending a fax with Fax Sr. is simple. Users can include cover page information and comments, and attach any documents in their native formats (such as .doc, or .xls) for faxing. Faxes are sent immediately, or scheduled, with complete fax status reported to the user. Even broadcast faxing is quick and easy.

SMTP Mail

Fax Sr.'s Internet capabilities extend to full support for SMTP mail systems. Users can send and receive faxes from any SMTP mail system exactly as they would send Email. From web browser mail systems such as Netscape Mail, faxing is automatic. Users can attach any documents or Drag and Drop files in their native formats.After the fax has been sent, receive a notification of fax status in the inbox.When a fax is received, the image is automatically placed in the SMTP inbox. By clicking on the attached image, you can view the fax on screen. Or you can print it, store it, forward, annotate or fax it to someone else.

Microsoft Outlook Web Access

Fax Sr. Intranet also supports Microsoft Outlook Web Access, the browser interface for Microsoft Outlook. Full integration delivers the ability to send faxes, receive faxes and be updated with fax status. Sending a fax is exactly the same as sending Email, including use of the Outlook Web Access phonebook.

Faxing Across the Internet

With Fax Sr., faxes can be routed across the Internet to the closest server to the fax destination.With Least Cost Routing and Global Routing, rules-based decisions allow you to find the optimal method to ensure faxes are delivered where and when you want, at the lowest cost.

Direct Internet faxing can reduce or eliminate long distance fax charges. Fax Sr. can link directly to Internet fax service providers such as PSInet, .comfax, and Ultranet to allow the Internet delivery of any fax. Any fax from any desktop is delivered by an Internet service provider, to a fax machine, automatically based on rules set by the administrator.

Windows NT and UNIX Servers

Fax Sr. supports both Microsoft's IIS and UNIX web server platforms such as Solaris. If a corporation has deployed multiple server platforms, Fax Sr. seamlessly integrates them all.

Feature Highlights

- ActiveX and JAVA. Fax Sr. includes both an ActiveX Web Client and a JAVA Web Client.
- Phonebook. Shared, public and private phonebooks. Or use the SMTP phonebook.
- Custom Cover Page. Use a standard cover page or design any number of custom cover pages.
- Notification. Automatic notification of your fax transmission, or receipt is displayed at the user desktop.
- Attach any Documents. Any type of document can be faxed, or attached to a message. Multiple documents can be attached in their native format.
- Broadcast. Broadcast to your mail lists or phone- book groups. Send immediately or schedule transmissions for later delivery.
- Inbound Faxes. Receive to Email inbox, Fax Sr. web client, and printer.
- No Client Software. No additional software is required at the client desktop.
- Multiple Simultaneous Email Systems. Multiple Email systems (MS Mail, cc:Mail, Exchange, SMTP) are supported on the same server.
- Least Cost Routing. Route faxes between servers, over the Internet or WAN to make free local calls out of a long distance fax connections.
- Security and Control. User name/password for log on security. Impose user access, time and region restrictions, or perform billback and user prioritization.

- Analysis. Powerful tools allow in-depth usage analysis, and graphical display by user or server from a complete audit trail. Full reporting capability delivers billback, usage, destination, and custom reports.
- No Dedicated or Proprietary Hardware. Fax Sr. is hardware independent and does not require a dedicated server.
- Scalable. As your need for a faxing system expands, Fax Sr. expands easily. Add users, servers, and lines easily to the Fax Sr. environment. Fax Sr. automatically performs outbound load balancing.
- Local and Remote Management. Monitor, configure, and control the entire Fax Sr. environment from any workstation. Receive unsolicited event and alarm notification. SNMP integration.

For more information visit: www.omtool.com

Chapter 27

Workgroup and Enterprise LanFax

with Optus FACSys

Today's IT departments face two fundamental challenges. First, corporate management is insisting that IT align itself more closely with the business. All technology investments must be aimed directly at critical business priorities such as building market share, improving customer "bonding," or reducing the cost of operations.

Secondly, IT must maximize the value realized from the various technology investments the business makes — which include systems, networks, applications and data. Shareholders are closely monitoring returns on capital, using next-generation metrics such as Economic Value Add. IT departments can't simply implement the current "technology du jour." They must take into account the real impact that each addition to the corporate technology portfolio will have on the return realized from all IT-related assets.

Based on these two objectives, Optus Software has taken an approach to network faxing that emphasizes unified messaging/workflow, enterprise integration, and reduced cost-of-ownership. By architecting fax messaging as a network service — rather than as a distinct application — Optus has succeeded in helping thousands of IT departments address their organization's primary business goals and extract significant additional value from its existing information assets.

Fax and unified messaging/workflow

Fax communications will continue to evolve as newer communications technologies such as Internet email and CTI become increasingly popular. It is certainly clear that both conventional and PC-based fax usage is continuing to rise. The universality, convenience and immediacy of fax is very attractive to business users. It's also a rich visual medium that allows graphics to be shared regardless of the format-of-origin. And there is a high comfort-level among business users because faxing is so simple and reliable. There are fax machines and fax-enabled devices everywhere — in hotels, airport lounges, delis, in homes in the form of PC fax modems, in cars in the form of PDAs. Like snail mail and face-to-face meetings, fax is a communications medium that is not going to go away.

What today's high-pressure information workers don't need is a fragmented messaging environment — where fax, email and voice are entirely separate from each other and require the end-user to constantly manage three different processes. Their productivity and ability to generate revenue is becoming increasingly contingent on fast, efficient, anywhere/anytime communications.

A well-integrated messaging platform can provide a wide range of business benefits. First of all, the consolidation of messaging media into a single platform eliminates the need to implement and maintain redundant and inefficient resources. Secondly, it simplifies life for the user. They don't have to check multiple in-boxes. They can send messages to a variety of different recipients without having to think about whether it's going via email or fax. They can forward messages received from one medium via another. They can archive and retrieve the different types of messages that they need to save for future reference. These benefits come about when IT enables users to treat messaging as a single application instead of as three or four.

Microsoft Exchange Server is a good example of a platform that provides an opportunity for integrating fax with other messaging types. Using Optus Software's FACSys Fax Connector for Exchange, IT can give users an integrated view of fax services. Phonebooks, the GUI, broadcasts and narrowcasts are all integrated across both fax and email, along with any other Connector type the customer chooses to implement. In addition to providing a unified approach for end-users to send and receive all message types, this approach also consolidates the administration of enterprise messaging resources and provides simplified access to what might be termed the enterprise "message warehouse."

The same holds true for IBM/Lotus Notes. Notes quickly becomes a very rich repository for vital corporate data — including documents, messages and information bases. Fax is important for sharing that data as required with parties-of-interest not directly wired into the Notes-based solution itself.

This is why it is critical that any corporate fax messaging solution fit into a larger strategy for unified messaging. Optus, for its part, is working with UM pioneers such as Octel and DEC/Mitel to seamlessly integrate fax into the corporate multimedia messaging environment. By integrating FACSys into Octel's Unified Messenger and DEC/Mitel's Media Path, Optus gives customers another means of exploiting the potential synergies between fax and voice.

For example, on the inbound side, FACSys and Unified Messenger allow users to have a single phone number for both voice and fax. The Octel platform recognizes fax transmissions and directs them to the fax server where they can be properly handled. That's a big step to making life simpler not only for corporate employees, but also for the people who have to communicate with those employees — including customers, suppliers and other partners.

While one of the primary functions of the enterprise message warehouse is to efficiently process various message types, it must also make those messages accessible to users in a variety of ways. In the Octel-Optus solution, users can access the fax server via what Octel terms its Telephone User Interface (TUI) to check on received faxes and forward them to any destination, such as a hotel fax machine.

In addition to email and voice, the other main medium of communications today is hard copy. There is a significant increase in the use of scanning to integrate paper documents into the enterprise information warehouse today. This clearly plays to the strength of fax as an image-friendly medium. Again, through strategic partnerships with companies such as Canon, Hewlett Packard, FileNet and Simplify — as well as through rich document format support and fax management features — Optus is ensuring IT managers of their ability to create complete end-to-end document workflow and communications solutions.

Extracting value through enterprise integration

Companies have a tremendous investment in the development and deployment of their mission-critical applications, as well as in the highly valuable data those applications house. Customer profiles, purchasing patterns, technical product information and other data types represent the bulk of a company's intellectual capital.

Fax can be very instrumental in squeezing additional value out of those knowledge assets. For example, a marketing department may want to do a fax "narrowcast" promoting a particular offer to a targeted prospect list. Or they may want to put technical documents at the fingertips of existing customers using a fax-back system. Or they may want to automatically trigger the transmission of a specially formatted fax document to a particular recipient as part of a business process.

Regardless of the specific type of faxing involved, it is essential for such an organization to have a well-integrated fax messaging server in place — one that can be fed any type of data from any type of system. Some companies run transaction processing on legacy systems. Others have midrange-based database applications. These same companies may have desktop-centric applications in place as well. Remote users also need a way to get to server-resident data and share it with others while at home or on the road.

To be able to "get to the fax server" from each of these points, it's vital that the corporate fax messaging solution provide an open, integration-friendly architecture. With FACSys Fax Messaging Gateway, Optus has provided a full range of access techniques, including:

Application-specific linkages — for popular enterprise products such as SAP R/3 and iXOS, Cardiff Teleform, Onyx Customer Center, Pivotal Relationship and PowerCerv's Application for Financials — allow users to leverage existing applications features

APIs — which facilitate the creation of scripted fax routines

DDE — for integration with standard Windows-based application development

Web Agent — to support intranet/Internet applications and remote access

Broad document format support — to allow the full range of application types to exploit the presence of the fax gateway, as well as to "feed" received faxes back into a full-range of corporate computing solutions.

By providing such diverse faxing methods, Optus ensures that users can employ a single, distributed fax messaging solution to meet all of its inbound and outbound fax communications requirements.

Managing the Real Cost of Fax Technology Ownership

All the benefits that integrated fax services provide won't deliver sufficient ROI for the corporation if those benefits are too expensive to achieve. IT decision-makers must therefore also take into account the life-cycle costs associated with the development, deployment, use and maintenance of an enterprise fax solution. Among the important factors to consider are:

Licensing Fax messaging applications are sold under a wide range of licensing schemes. While per-channel pricing structures may initially look more attractive, the key phrase is "caveat emptor." Fax channels can quickly become saturated by new fax applications, ultimately yielding a very low ratio of actual users to channels. Hoped-for savings typically evaporate within a few quarters of the initial implementation. With per-user pricing, this down side risk is eliminated and IT can be sure of fixed software costs regardless of fax volume.

Least Cost Routing

An additional benefit of integrating fax with an enterprise messaging platform such as Exchange is that you can leverage the corporate WAN/mail infrastructure to implement Least Cost Routing. By telling Exchange's routing facilities to send faxes destined for specific area codes from specific Exchange servers, organizations can significantly reduce long distance charges. So if someone in New York is sending a fax to someone in LA, and the company has an office in LA, Exchange will pick up the 213 area code, select the server in LA as the point-of-transmission and send it from there. For multi-national firms, this type of LCR implementation can yield even more substantial savings.

Advanced cost routing capability is also available to FACSys users through Optus Software's partnerships with major, worldwide telecom companies and service providers such as MCI, Cable & Wireless and Xpedite that have a global infrastructure in place for sophisticated movement of messages and documents. And, Optus continues to release progressive products which provide for integration of rules-based message delivery with other Workflow Partner products to ensure crucial faxes are sent in the most timely and cost efficient manner possible.

For more information, visit: www.facsys.com

Chapter 28

Work Group and Enterprise LANFax

with RightFax Enterprise

Introduction

Faxing is the lowest common denominator in business communications. Almost everyone has at least one phone line and far more people have fax machines than email or voice mail. However, email and voice mail are growing in popularity and users are searching for ways to consolidate their messages. At the same time, sending and receives faxes over the public switched telephone network is quickly giving way to Internet and Intranet routing due to the significant cost savings. RightFAX fax server technology has set the pace in these advancements. Our document-centric society will never allow fax to be replaced by email or voice mail, however there are ways to integrate messaging systems and enhancing fax. These enhancements have led to the development of a new generation of fax servers—the enterprise fax server.

Fax Server Market

Market Segmentation

The network fax server market encompasses a diverse group of industries and business segments, each with their own specific needs. RightFAX has categorized these needs to create specialized products for the market.

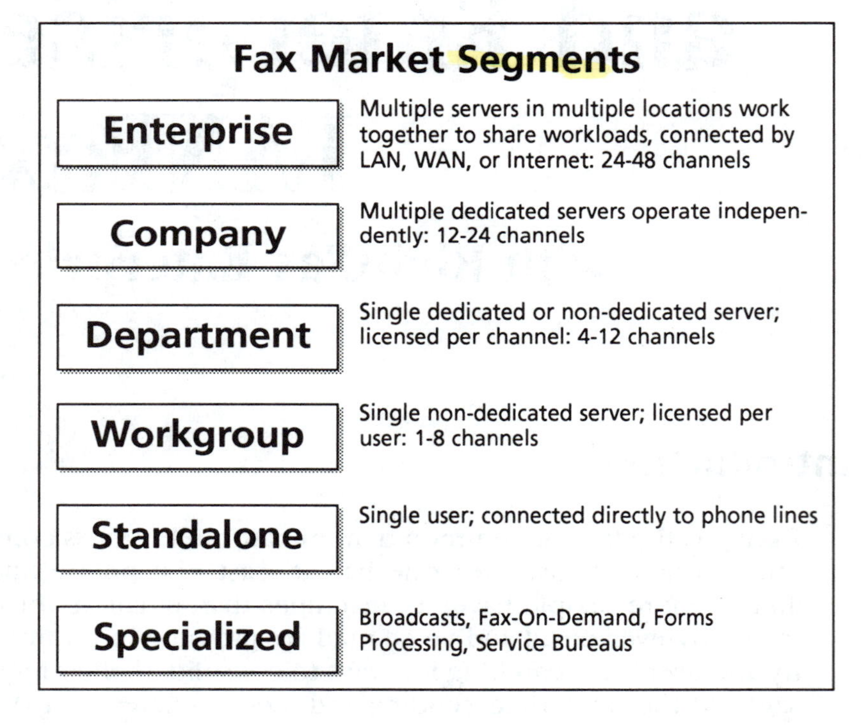

Enterprise vs. Workgroup

The advent of the Internet has fueled the demand for collaborative communications. Five years ago, more fax servers where being sold in the Workgroup and Department markets where servers work independently. Today, organizations have developed their networking infrastructure and require large roll-outs of networking software in multiple sites across the enterprise. The maximum benefit comes from allowing these servers to talk to each other and share resources.

Enterprise Technology

What makes a fax server an "enterprise fax server"? Choices.The more sites, servers and users you add to the network, the more flexible your faxing system must be.The last thing any administrator wants is to make decisions today that limit his options tomorrow.

When searching for an enterprise fax server, several issues should be evaluated:

- How and what type of faxes is my business sending and receiving?
- What do users and administrators expect from a fax server?
- How scaleable is the fax server?
- Will it integrate with my other network applications?

RightFAX, Inc. designed its fax server software specifically for the enterprise. Of course, an "enterprise" consists of many different sites, each with a different size, purpose and expectation. RightFAX provides hundreds of options to help meet the needs of every enterprise, regardless of size, faxing volume or type of faxing. Many of those options leverage Internet, WAN and LAN connectivity to help reduce faxing costs, increase productivity, unify communications and integrate with the rest of your network.

Server-Side Strength

Architecture Overview

The RightFAX enterprise system is composed of four main parts: the fax database and images, the fax server programs, the administrative utilities and the client software. RightFAX's 32-bit client/server architecture is unique.All fax information is stored in a central database on the fax server. This allows users to access the same fax information regardless of how or where they choose to log into the fax server. For instance, user Bob may open FaxUtil (RightFAX's Windows client) and view his faxes. Then a few hours later, he may view the exact same fax list by logging into the RightFAX Web Client. The fax list, phone books, and other data will always appear the same across any platform, via any client interface.

The RightFAX Way

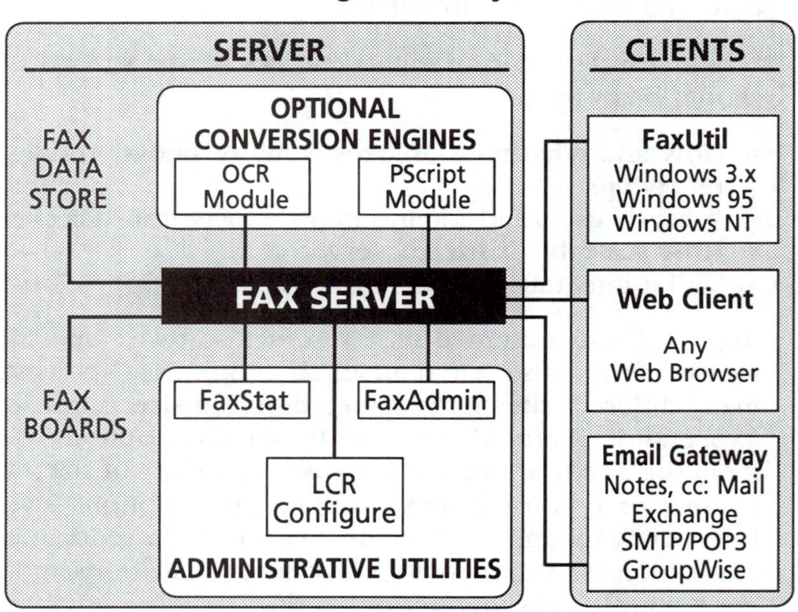

Other fax servers do not store fax data in the same way. Many store a copy of the fax data on the local hard drive. The server only stores the new fax data (that which has not yet been copied to the local machine). This means that if user Bob accesses his faxes through a Windows client, then later logs into a Web client to view his faxes, he will not see the same fax information. He will only be able to access the new fax through the Web client. This process severely limits the flexibility of the fax server. RightFAX ensures that you will always get the same information and the same functionality, no matter how you access your faxes.

The Wrong Way

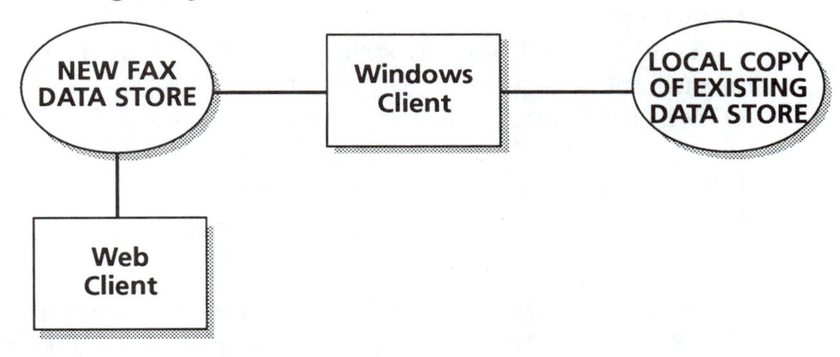

Fax Delivery

Delivering faxes in the fastest, least expensive manner possible is one of the biggest challenges facing fax administrators. Here again, making a bad decision today might limit future options. RightFAX does not limit your fax delivery options. Whether you choose to send and receive faxes via the Internet, the LAN, the WAN or the public phone lines, RightFAX provides the flexibility you need.

Public Switched Telephone Network (PSTN)

Each enterprise fax server should allow you to connect at least 48 phone lines. RightFAX supports varying combinations of analog and digital lines, including loop-start, DID, T1, E1 and ISDN lines. The PSTN is more expensive for long distance and international calls as compared to alternative routing methods. However, for local calls, the PSTN is the best delivery method due to the far-reaching presence of telephone lines.

Deciding the best way to route faxes coming in from the PSTN to users in your organization takes some time and careful evaluation. Routing method decisions influence decisions about phone lines and fax boards types. RightFAX lets you choose from among the following inbound routing options:

- DID/DNIS (Direct Inward Dial/ Dialed Number Identification Service)
- CSID/TTI (Caller Subscriber ID/Terminal Transmitter Identifier)
- DTMF (Dual-tone Multi-frequency)
- ISDN (Integrated Services Digital Network)
- OCR (Optical Character Recognition)
- Line or Channel
- Manual

Of all these methods RightFAX most often recommends DID or DNIS (digital DID) routing, because it is the most economical and efficient. It allows a phone number to be assigned to each recipient, while not requiring a physical phone line for each number. Phone number are purchased in banks (e.g. 320-7000 through 320-9000). Users are assigned routing which match the digits passed down from the phone line to the fax server. Faxes are then routed directly to the recipient. It's accurate 100% of the time and it does not require the sender to input any other information than the fax number.

Other methods like CSID and OCR are less accurate, simply because the technology is less precise. Routing faxes manually or by fax line can prove expensive because of the investment in manual labor or physical phone lines. Routing by DTMF requires input of a fax extension in addition to the fax number, so accuracy depends largely on the sender.

Internet/Intranet Fax

As an alternative to the PSTN, fax servers can route faxes over existing network connections such as the LAN, WAN and Internet. RightFAX enterprise servers leverage this connectivity automatically with its least cost routing technology.

Intelligent Least Cost Routing

It's no secret that least cost routing is the computer-based fax industry's hottest technology. Everyone wants to buy a fax server that promises to save their business money. Identifying this need in your organization is the easy part. The hard part begins when you try to evaluate and decide which fax server best meets your needs today and tomorrow.

What is Least Cost Routing?

Least cost routing (LCR) is a term used in the network industry to describe technology that allows you to route data using the least expensive path. The fax industry has adapted LCR to include the process of routing faxes among fax servers on a network. In either case, the goal is the same: to save money by reducing traffic over the public phone lines where ever possible. By using an organization's Internet or existing Intranet connections, faxes can be routed to the server closest to the fax destination, potentially reducing a fax from a long distance call to a local call.

Simple Least Cost Routing

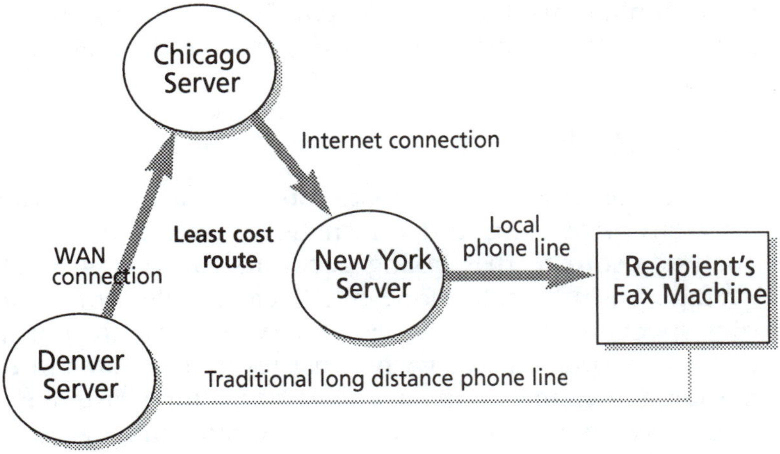

The term "least cost routing" is somewhat confusing. Fax servers do not actually make decisions based on price structures or tariff algorithms. It is up to the fax administrator to create rules that appropriately reflect the company's faxing habits and phone carrier rates.

Intelligent Least Cost Routing vs. Simple LCR

The challenge of instituting LCR for the entire enterprise lies in the limitations of most fax servers. Administrators should be able to create routing rules that take into account several factors, such as time of day, fax priority, server availability and server workload, in addition to matching the number dialed. At the same time, these complex plans should be flexible enough that the administrator need not create individual rules for every combination of numbers. Perhaps most importantly, administrators want intuitive tools to help them manage and troubleshoot such a system.

The needs of the users are simple. They simply want the fax to make it to its destination in a timely manner. They do not care how it gets there or whether it is routed by LCR. LCR should be transparent to users and they should not have to change the way they enter fax numbers.

Most fax servers, with simple LCR systems, cannot overcome these limitations. Only *Intelligent LCR* from RightFAX can give users and administrators the power and flexibility they need.

Intelligent LCR Benefits

If your business is like most, you send a combination of intrastate, interstate and international faxes. In fact, some analysts estimate that faxing accounts for over 40% of the average business' telephone bill. Because telephone companies specialize in many types of services, you are not likely to find any one carrier with consistently low rates in every market. For example, an east coast-based long distance company may offer competitive intrastate rates in New York, but offers higher rates on calls between New York and Washington, D.C. That same carrier might offer competitive rates to England and France, but higher rates to Japan.

Searching for and keeping track of the "best value" can require a lot of effort, but that information, when used in conjunction with even simple LCR systems, can pay high dividends. When you're sending high volumes of faxes every day, saving only a few cents per page can add up to a significantly lower phone bill at the end of each month. The basic features of LCR include:

- Reduction of long distance charges
- Use of the Internet as an alternative to the public switched telephone network (PSTN)
- Full preservation of sending options (cover sheets, automatic printing, billing codes)
- Transparency to users—they don't need to adjust the way they enter fax numbers

RightFAX's Intelligent least cost routing extends the functionality of basic LCR systems, offering even more sophisticated features and options for even greater returns. In addition to the cost savings, you can leverage the power of multiple servers on your network, increasing server efficiency by sharing resources and workloads. Instead of investing in a 24-channel server at every site in your organization simply to ensure you have enough power at peak times, you can purchase more fax channels for sites that are busier and fewer channels for sites that are typically less busy. Then, when you need that extra bandwidth, you can distribute faxes to an available server. In short, your servers are more productive with less overhead invested. Only RightFAX fax servers offer these Intelligent LCR benefits:

- Capacity sharing (load balancing)
- Dynamic capacity tuning
- Priority, time, and day restrictions
- Cascading route combinations
- Integration with service bureaus (Enhanced Fax Service Providers)

- Automatic error recognition and immediate rerouting
- centralized management and testing tools

Intelligent LCR satisfies everyone's needs. Management will see the dramatic reduction in long distance charges. Administrators will enjoy the intuitive, flexible power of Intelligent LCR systems. Users will not need to be retrained on how to fax, because nothing on the client side changes. Intelligent LCR pays for itself in more ways than one, in just a few short months.

Example of Intelligent Least Cost Routing for a Distributed Company

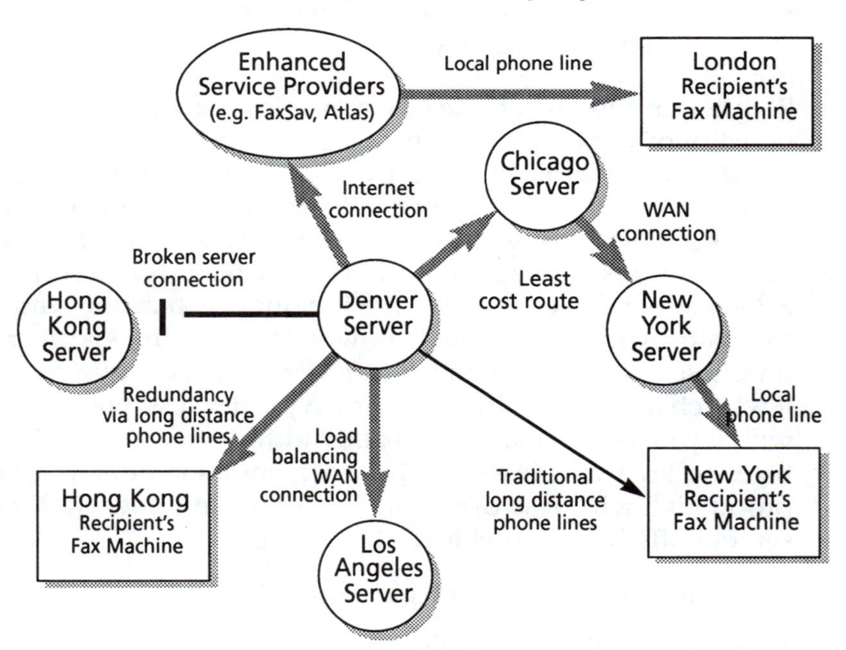

How does Intelligent LCR work?

Intelligent LCR is comprised of several components, including: dialing plans, testing tools, and management utilities. Each is described in detail below.

Dialing Plan

Dialing plans allow the fax server to make sophisticated decisions about where a fax is going based on the phone number, time of day, day of week, fax priority and fax source. The dialing plan is a key component and is actually useful even on single fax server installations.

The fax server's dialing plan is made up of several dialing rules. Most fax servers compare the fax number entered to the list of dialing rules and stop when they find the first match. This first match may or may not be the "best" match and the fax may be routed incorrectly. Because the dialing plan only allows one rule to match a number, the administrator faces the difficult task of creating individual rules that match every possible scenario for each number.

Intelligent LCR saves the administrator from trying to anticipate every combination entered by users. Using the defined rules, the server evaluates each fax number entered by users on your network. After comparing the fax number and conditions (time, day, priority) to each rule, a "weight" or measurement is assigned. A number can match more than one rule at a time. The more closely the number matches a rule, the higher the weighted value it is assigned by the server. The fax server will use the rule with the highest weight to send the fax.

Pattern Matching

While most fax servers' simple LCR systems may allow simple pattern matching of phone numbers, Intelligent LCR allows complex pattern matching. For example, a product may only allow a single, match-everything wildcard character (e.g. 415+). Intelligent LCR has four different wildcards that allow ultimate flexibility (e.g. 415+, 415???????, 6?5+, ~415+, ????? and more).

Prefix Tables

Intelligent LCR allows the administrator to create prefix

tables to store and organize groups of numbers such as prefixes or area codes. Rather than creating a separate rule for each number pattern, the administrator would simply create a few rules that reference this prefix table, saving considerable time without compromising functionality.

Time Sensitivity

The Intelligent LCR dialing plan can also be tailored by using time-of-day and day-of-week settings. That is, a dialing rule may only be effective during certain times of the day and certain days of the week. During the rest of the time, the rule is out of contention.

Priority Sensitivity

Intelligent LCR's sophisticated dialing plan allows the system administrator to define rules stating that faxes of certain priority levels can be routed via specific servers. Now, when a user sends a high priority fax, it will always go out via the local phone lines. Most company would agree that any cost savings of LCR would be easily lost if the fax were not delivered due to a WAN link or remote server failure. To put it another way, saving 50 cents while sending a customer a $50,000 proposal is considered too risky.

Priority sensitive rules would also be useful if a company had scheduled a low priority fax broadcast to 20,000 people and didn't want to "starve" their normal and high priority faxes from being sent. Intelligent LCR would route the broadcast to one specific server, leaving the other remote servers and the local server available for the rest of the fax traffic.

Redundancy

Now that faxes may be sent through physically separate servers, Intelligent LCR must take extra precautions to ensure that each and every fax is successfully delivered in a timely manner. To accomplish this, the server makes some intelligent decisions when it detects faults. There are two likely problems encountered when using LCR: WAN link failure or remote server failure.

LCR When Connections Fail

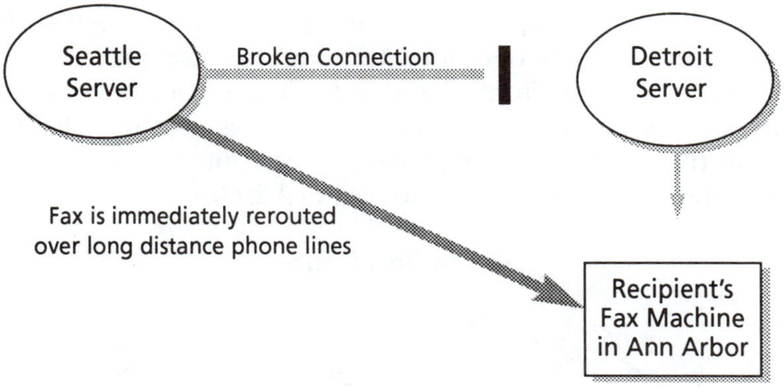

Seattle Server

Broken Connection

Detroit Server

Fax is immediately rerouted over long distance phone lines

Recipient's Fax Machine in Ann Arbor

Company XYZ creates a rule that routes a fax from the Seattle server via a WAN link to the Detroit server. After the fax is successfully scheduled for transmission in Detroit by the Seattle server, the WAN link fails. The Seattle server now has no way of confirming successful transmission of the fax on the Detroit server.

Traditional fax servers take a simplistic approach to such conditions - they simply wait up to 8 hours (or some definable delay) for confirmation from the Detroit server. Only after the confirmation does not arrive would the Seattle server attempt to retry the sending operation. Users will not tolerate such lengthy delays, as it is almost easier to mail the document than fax it.

Intelligent LCR will detect the WAN link failure almost immediately (within minutes). Upon doing so, it will temporarily disable all dialing rules which would direct a fax to the downed server. All faxes are now processed as if the path to the downed server doesn't exist. If alternate routes exist, those are automatically used according to weights and server loads. If no alternate routes exist, the local fax server sends the fax directly over regular phone lines. Users and administrators can rely on our philosophy that "the fax must go through."

Cascading Routes

With simple fax servers, setting up routes between three or more fax servers becomes tedious due the fact that these servers force the administrator to define routes for all possible combinations of servers. For example, in a five server system, there are ten combinations of routes which must be maintained as depicted in diagram A below.

Cascading Routes

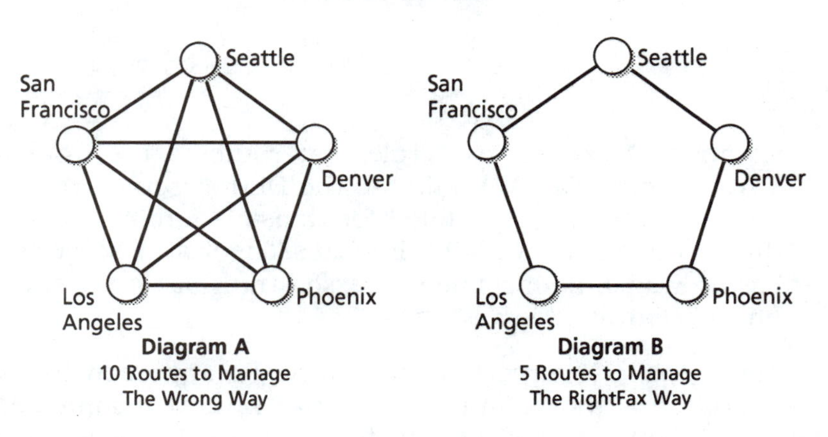

To avoid the problem of exponentially growing route combinations, Intelligent LCR uses cascading routes as shown in diagram B. That is, Intelligent LCR servers can re-route faxes which have been scheduled from remote servers.

Load Balancing

Load balancing allows multiple servers to share workloads for outbound faxes. Intelligent LCR accomplished this money-saving feature by allowing the definition of multiple destination servers in the dialing plan and allowing more than one rule to match a number. This means that if the local server is busy (e.g. all its channels are currently sending or receiving faxes), it can route faxes to another, less busy server to be sent immediately. Because traditional LCR systems can match only one rule per number, there is no way to balance workloads. With simple LCR, overhead costs increase because more fax lines are required and server efficiency is decreased.

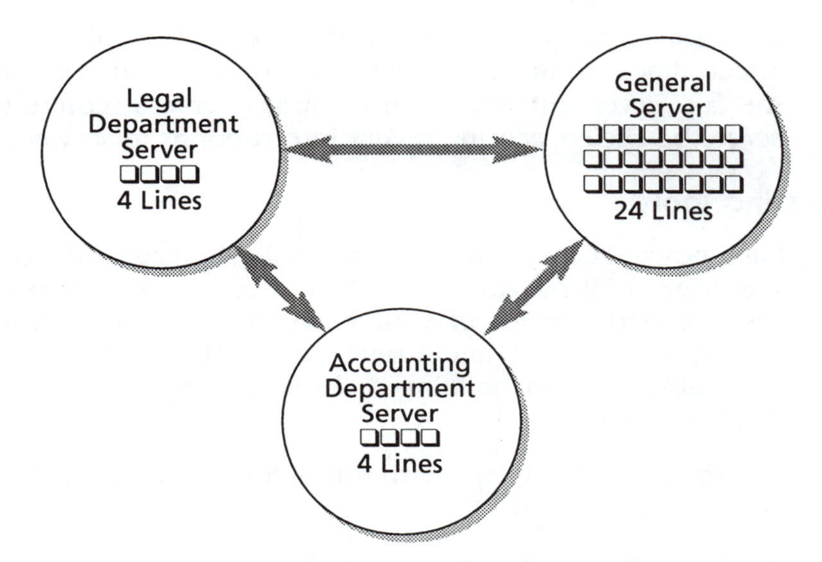

When multiple rules match a phone number with **equal** weight, Intelligent LCR will direct the fax to the server with the most available phone lines. If both servers have all outbound fax lines in use, then the fax will be directed to the server that will have a free phone line in the shortest amount of time. For example, Server1 and Server 2 both have 10 outbound phone lines and all lines are busy. However, Server1 has 200 faxes in the queue where Server2 only has 10 faxes in the queue. Server2 will likely have an available phone line before Server1, so the server directs the fax to Server2.

Central Management

Multiple, physically-separate servers can be difficult to manage. To alleviate this problem, Intelligent LCR provides administrative tools for Windows 95 and Windows NT workstations to allow an administrator to manage all fax servers from a central location. Communication can occur using IPX/SPX or TCP/IP protocol sets over the LAN, WAN, remote access link

(RAS), or even the Internet. So, an administrator in Phoenix can configure and revise dialing plans for San Francisco, Seattle, and other servers without leaving his office.

Changes to server configurations and dialing plans take affect almost immediately without stopping and restarting the fax server software. Competing fax servers require the server be cycled, causing severe interruptions in service.

Testing Tools

Such power LCR systems require intelligent diagnostic tools. Intelligent LCR provides several advanced tools to help you test the performance of your enterprise fax server: Route Tracer, Server Ping, and Route Tester. These tools prove invaluable tools in building and maintaining the complex dialing system.

Route Tracer traces the path through which a fax might be sent, detailing:

- Each server a fax passes through;
- Which rules are being used;
- How the phone number is manipulated along the way;
- Hop counts; and
- Hop timings.

Route tracer helps the administrator test the dialing rules he has constructed with real fax numbers without actually sending a fax. This proves invaluable in diagnosing unexpected fax routes.

Server Ping will test communications with a server over a particular protocol, times the round trip of such packets, and shows current server loads. This feature helps the administrator determine the servers availability and perhaps redistribute workloads accordingly.

Rule Execution determines why a fax would or would not use each rule in a dialing plan. When multiple rules match a particular phone number, Rule Execution shows each one and the weight assigned to each match. This tool helps the administrator to determine why a particular number followed a particular rule and whether the rules need adjustment.

Show Downed Servers displays all rules which are currently disabled due to communication errors between servers. The error which caused the rule to be disabled is also displayed, allowing the administrator to easily diagnosis the problem.

Enhanced Fax Service Providers (EFSP)

Intelligent LCR offers the unique ability to route faxes to external service bureaus, also known as Enhanced Fax Service Providers (EFSP). Businesses such as FaxSav and Atlas Telecom (AFAX) offer high throughput, guarantees of delivery, high quality, and reduced costs as compared to the public switched telephone network. An EFSP can usually deliver thousands of faxes in a matter of minutes or hours because they have hundreds or thousands of fax ports in concurrent operation.

Additionally, EFSPs have developed a large, global network. In many cases, their infrastructure extends to thousands of sites. Many companies rely on their services because it is not possible or feasible to have an internally owned and operated server in every country, state, or city to which they send faxes. EFSPs are able to offer a reasonable alternative to such businesses.

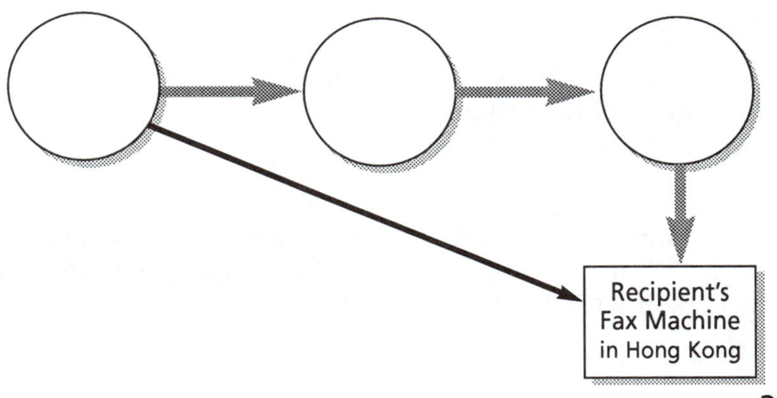

LCR via Enhanced Fax Service Providers

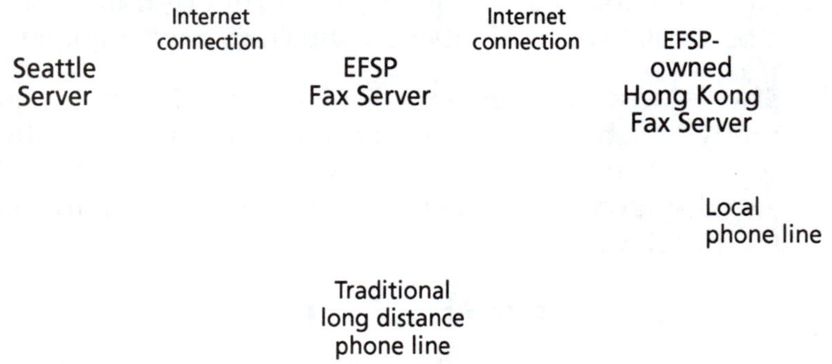

Seattle Server

Internet connection

EFSP Fax Server

Internet connection

EFSP-owned Hong Kong Fax Server

Local phone line

Traditional long distance phone line

The server side of fax server software lays the foundation for the performance of the application. If the server is not built to be scaleable, flexible and reliable, users and administrators alike will be disappointed. RightFAX servers are developed to be strong and dependable. At the same time, the server and database architecture allows RightFAX to integrate with the latest technologies, improving fax server functionality.

Client-Side Flexibility

One of the most basic requirements of enterprise fax servers is the ability to send and receive faxes quickly and efficiently. Users want to be able to fax from whatever machine they are using, where ever they are using it. RightFAX developed several ways to access fax information and exceed users' expectations:

- Windows clients for the LAN
- RightFAX Web Client
- Email Connectivity
- Host Integration

Communication with the fax server can occur using IPX/SPX or TCP/IP protocol sets over the LAN, WAN, remote access link (RAS) or even the Internet.

LAN Client Utilities

RightFAX developed FaxUtil, a user interface, and FaxAdmin, an administrative interface for easy management of faxes and users. These clients were designed to run over the local area network on Windows 3.x, Windows 95 and Windows NT workstations.

FaxAdmin was created to help administrators easy create and maintain fax objects such as users, groups, library documents, forms, signatures and printers.

FaxUtil was created to provide users with a central fax mailbox from which they could send, receive, view, print, forward, route and annotate their faxes. Users could create a fax on-the-fly or print a fax from another application (e.g. Microsoft Word). After completing necessary cover sheet information such as the recipient's name and number, the fax is sent.

Received faxes can be routed directly into the recipient's fax mailbox. RightFAX also includes an automatic fax distribution feature which sorts faxes among a group of users in a linear or round-robin fashion.

The LAN client utilities also include the RightFAX PowerBar, RightFAX tray icon, and Quick Fax options. These tools are shortcuts that allow you to switch fax print drivers, attach library documents, or send broadcast faxes quickly and easily.

Email Connectivity

The move toward a universal inbox relies upon email connectivity. Users are familiar with the process of creating and sending email messages. In addition, many users are able to access their email messages remotely over the Internet, when they are out of the office. The RightFAX Email Gateway provides the necessary fax-to-email connectivity and supports the following email systems:

- Lotus Notes
- Lotus cc:Mail
- Novell GroupWise

- Microsoft Mail
- Microsoft Exchange
- SMTP/POP3 Internet Mail

When sending a fax from email, users specify the recipient's fax addressing information in place of an email address. The message is then sent to the RightFAX Email Gateway, which is usually set up to look like another post office on your mail system. Anything addressed as a fax is routed through the main RightFAX system which converts it to fax format and transmits it over the phone lines, WAN or Internet.

Received faxes come in via phone lines or other RightFAX servers. RightFAX knows the mail account to deliver it to by the routing code assigned to the fax. Fax messages can be copied or moved entirely into the email system in a variety of file formats.

The RightFAX Email Gateway lets the fax server leverage the conveniences of email messaging and add the benefits of fax messaging. It provides seamless communications through one interface, connecting everyone in the organization, regardless of workstation type or location.

Web Client

The RightFAX Web Client is simply another means by which to access fax information in the RightFAX database. It is similar to FaxUtil, RightFAX's other user interface, in that the user makes requests to the fax server and the information returned is displayed graphically.

There is no software to install on the user's workstation. Simply open a World Wide Web browser and enter the URL (web address) of your fax server.

FaxUtil communicates directly with the RightFAX fax server to get information. The RightFAX Web Client must use a different method to communicate with the RightFAX fax server due to Internet protocols. Through the Web browser (i.e. Netscape or Internet Explorer), the user makes requests for information from the fax server.The request travels from the browser to the Web server which then translates and delivers the request to the RightFAX fax server. The fax server asks the RightFAX database for the answer and returns it via the Web server to the Web browser.

The RightFAX Web Client runs over a Secure Socket Layer (SSL v2 or v3) which is a security protocol that is included in Microsoft's Web server. Basically SSL ensures that any data transferred over the Internet is encrypted so that no one can view or intercept it and read the information. Of course, the RightFAX fax server's security is not compromised in any way.

The RightFAX Web Client adds unparalleled advantages to the RightFAX fax server without compromising fax quality, speed, or security.

Database Connectivity

Most organizations have a centralized database of contact information. RightFAX can connect to those databases and retrieve information if the database is ODBC- or MAPI-compliant. Available on Windows NT and Windows 95 clients, users are able to search, sort and select one or more entries from their Exchange Global Address Books, Outlook Contact Manager, SQL Server, dBase, Access or other database for single faxing or fax broadcasting.

If the organization does not have or chooses not to use an ODBC or MAPI contact database, users can store their fax address information in a separate phone book in RightFAX.

Automatic Fax Rules

You can set rules to automatically forward every fax you receive to a fax machine or another user on the network. In addition, incoming faxes can be automatically OCR'ed and delivered to your FaxUtil mailbox or email system. You can also automatically print all sent or received faxes to any printer on the network. RightFAX can also be configured to notify users in a variety of methods, including by network broadcast, email, voice mail, and more.

Administrative Control

The sophisticated nature of fax servers demands comprehensive management utilities. RightFAX provides centralized management tools that can be executed from any Windows NT or Windows 95 workstation on the network. You do not need to be at the server to dynamically reconfigure it or simply check status and performance information.

Administrators can also easily import existing network users into RightFAX and automatically assign routing codes for quick and easy setup. RightFAX allows fax administrators to be configured on an server-wide or group-wide basis for even greater flexibility.

File Format Flexibility

Along with the ability to send and receive faxes from any platform, anywhere in the world, users expect flexibility when it comes to fax formats. In addition to sending standard fax formats like TIFF-G3, the super-compressed TIFF-G4, PCX, DCX and ASCII/Text formats, RightFAX will convert documents attached in their native file formats. Over 45 of the most popular Macintosh and Windows file formats are supported, including MS Word, HTML, JPEG, GIF, WordPerfect, PowerPoint and more.

Received faxes can be routed to network directories, email or voice mail in a variety of formats including TIFF-G3, TIFF-G4, PCX, DCX, or ASCII/Text.

Reporting and Cost Recovery

RightFAX offers the most robust cost recovery tools in the industry. Users can be required to enter up to two billing codes that can be validated in several ways for each fax sent or received. You can also use the unique ID automatically assigned to each fax for billing purposes. If you prefer, you can create a table of phone codes on the fax server and let RightFAX automatically assign billing information on a user-basis. Then, to gather the data, use our built-in reporting systems or export fax history information to your own accounting system.

Integration and Customization

RightFAX provides easy-to-use embedded codes to help automate faxing. These codes, which are placed in your fax documents, give RightFAX the cover sheet and attachment information necessary to send your faxes to their destinations. In addition, embedded codes can be used to adjust the way the server sends the fax or faxes on a case-by-case basis. Embedded codes are compatible with DOS and Windows applications, as well as Unix, AS/400 and other host platforms. They prove extremely useful for broadcasting faxes to large groups of recipients and for writing simple integration between RightFAX and other software.

Additionally, RightFAX has forged partnerships with manufactures of leading workflow processing, document management, cost recovery, forms processing, voice mail and other business software. These partnerships have resulted in native integration and integration kits for a variety of complementary technologies.

Making it Work for You

The era of the enterprise fax server has arrived. As the

Internet and Intranets allows for more and more collaboration between sites and departments, the enterprise fax server unifies business communications. Now you know that all fax servers are not created equal. RightFAX has a head-start on enterprise fax servers, developing and implementing technology that truly saves businesses time and money without compromising flexibility and control.

RightFAX satisfies everyone's needs: administrators and users will appreciate the power and simplicity of its design; management will appreciate the bottom line. It's technology your enterprise cannot afford to be without.

For more information visit: www.rightfax.com

Chapter 29

Enterprise and WANFax

with Interstar Technologies

Interstar Technologies Inc designs and develops fax server software for organizations seeking enterprise-wide solutions. Interstar has fax servers and clients on two platforms: Faxtrek and Lightning FAX (for AIX); and Lightning FAX (for Windows NT). They also developed customizable application modules that may be used in conjunction with servers to meet the needs of call centers, and email to fax gateway users. Finally, their integration toolkit extends fax services to non-Windows computer systems.

Interstar's Approach to CBF Technology

The philosophy behind FaxTrek, was to integrate the best technology available in offering a product that could be relied upon to perform mission-critical fax communications.

When they started, the Network Operation System (NOS) of choice was AIX, the commercial version of UNIX developed by IBM. Its market share, scalability, and fault tolerance features were well suited to meet their enterprise solution criteria. Efficiency was to be achieved through the use of superior hardware and by the power and stability offered by the NOS.

Interstar chose to develop the product around Dialogic's GammaLink boards. Choosing fax boards over modems enabled them to offer a tremendous throughput without taxing the server's resources. Since much of the processing is done on the board itself, a single server can accommodate up to 720 lines (shortly 1440).

FaxTrek for AIX

Features:

NOS and hardware

- RS/6000-based AIX Fax Server
- Support for both Micro Channel and ISA Bus architecture
- Scaleable up to 48 channels per chassis (soon to be 96 channels)
- Supports all GammaLink, Dialogic and Dynatel boards providing robust high fax volume capability
- OSF/MOTIF, ASCII Terminal mode environment.

Fax Management

- Incoming and Outgoing FaxLists and Archives
- View or print contents of any fax from the FaxList
- View transmission details and monitor activity
- Delete either listings or actual fax files

Phone books

- Multiple phone book capability
- Save Phone Books in *.dbf format and ODBC interface
- Export phone books to ASCII format
- Print Phone Books or view and edit them on screen.

Cover Sheet Editor

- Drawing and annotation capabilities
- Supports the importing of TIFF 5.0, PCX and BMP files
- Zoom features: 25%, 50%, 100% and 200% views
- Customizable variable fields that automatically insert Phone Book data into your cover sheets.

Viewer

- Fully compatible with all Group 3 and 4 TIFF files
- Same zoom features as Cover Sheet Editor
- View thumbnails of entire documents
- Deleted unwanted pages.
- Fax Annotation

Other important features

- Automatic fax printing
- PCL5, Postscript conversion
- MS-Windows Explorer workstation interface
- Attach sounds to specific fax events under Windows
- Fax broadcasting
- DID, ANS/DNIS and DTMF support to route faxes to specific users on a network
- TrueType font support with AGFA Font library included.

Reliability and Savings Without Pain

One of FaxTrek's main management tools is the Fax Monitor which provides real-time, graphical information reporting (figure 1) of the system's fax activity.

Reports of individual fax board send/receive status, global fax activity percentage, and other detailed processing information, allows organizations to monitor and maximize their system's efficiency.

Businesses that regularly send and receive faxes can increase productivity while saving time and money in the following areas:

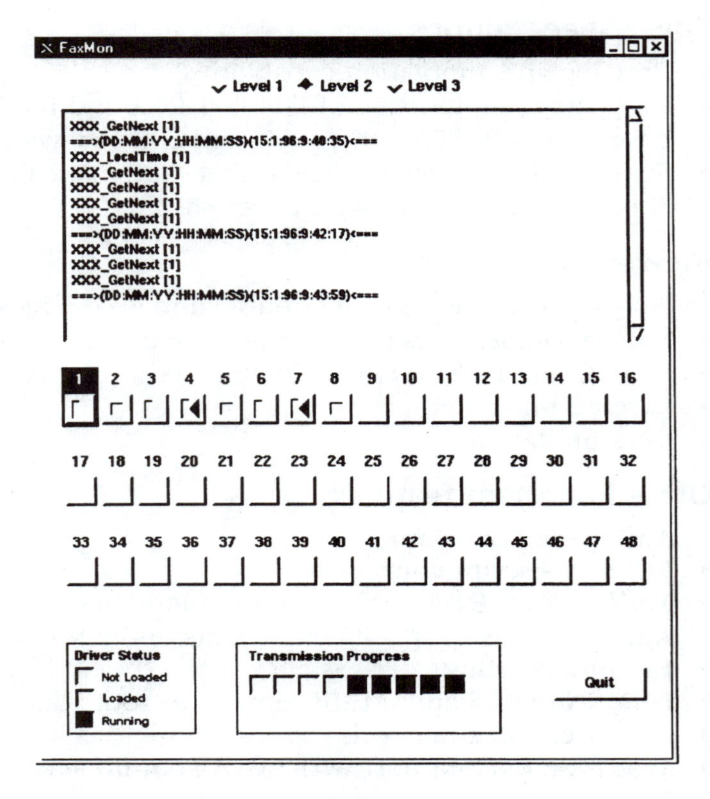

Figure 1

Communication Costs

Due to the performance of the GammaLink board, FaxTrek offers increased savings on phone charges due to the board's sophisticated compression capabilities, and greater connectivity as a result of the its wider range of faxing protocols. In turn this yields a higher rate of fax transmissions. In addition, FaxTrek has full service support for incorporating the use of T-1 and E-1 lines and also of ISDN (BRI and PRI), often required by large installations.

Productivity

Using the appropriate fax server is like installing a personal fax machine for each employee. If your organization has not converted to CBF, consider the time and energy that are wasted by your staff just waiting around fax machines. All this could be a thing of the past with the right CBF system. Information that is not acted upon on in time may be wasted. With the right routing protocol (DID, ANS/DNIS and DTMF), all your employees could receive, directly to their workstation, all the faxes that are addressed to them. Nothing gets lost, all critical documents are received at once.

Security and confidentiality are primary concerns for any organizations. Primarily, all log-in procedures are password-protected to secure unintended users from gaining access to restricted faxes. User Access Rights are defined in three different categories: Administrator, Super User and User. This hierarchy permits a high degree of security while allowing flexibility.

Administrators control and define the majority of privileges and rights of other users on the network. However, certain options, such as layout of the Fax List, open for customization by each user. The Fax list helps users and administrators alike to manage even high volume fax traffic. Visual symbols identify, at a glance, transmission types and status. Further details of each fax are only a mouse-click away.

FaxTrek supports TrueType font technology, with conversion to both PostScript and PCL 5 printer languages. Plus, a special AGFA font package is bundled with every solution.

FaxTrek's full-featured Fax viewer displays every detail of each fax document with the view you want (25%, 50%, 100% or 200% magnification); or users can use the Thumbnail mode to view multi-page documents in a convenient screen format. Once a fax is opened in the Viewer, either text or graphics can be applied with the intermediate options of retransmittal or printing.

Today's enterprise-wide solutions must be scalable. Through its Application Programming Interface (API), FaxTrek was designed to allow other applications to send information via fax without having to engage a separate fax interface.

This allows enterprises to incorporate fax services in their automated data processing routines in order to boost productivity.

Sending Faxes:

The Fax Queue module allows you to manage all incoming and outgoing faxes. It does this by showing you the list of all your faxes and providing you with the following options for managing the list.

View: view the list of incoming and outgoing faxes, including information about their status and identity.

Open: open TIFF and ASCII text faxes to see their contents.

Print: print a hard copy of any fax in the list.

Delete: delete any selected fax or faxes from your system.

Re-Submit: You may resubmit any fax for transmission, whether they were successfully transmitted initially or not.

Route: This feature allows you to re-send a selected fax to another user on your network.

Clean: Clean your Fax List by compressing the database.

Viewing Faxes

FaxTrek makes viewing your received or sent faxes very easy. Viewing your faxes can be done in two ways. Click either on the Outgoing Fax List button or on the Incoming Fax List one. The Fax list will be displayed.

Routing Faxes

FaxTrek allows a user to route (forward) a fax to another user on the same network. When routing a fax, you have the option of moving the faxes to the selected user or just routing copies.

Case Study

AUS-based rail transportation company with more than 40,000 employees and operations in 22 States and 2 Canadian Provinces was looking for an alternative to their existing fax server installations. Their existing servers, from various manufactures, proved unable to cope with the increased traffic demands on their network, thus a new product was sought.

The Challenge

The task at hand was to seek a CBF solution for the transmission of itineraries that had to be faxed out concurrently, overnight, in various regions across the continent. The project mandated that the system must be powerful enough to send over 11,000 faxes within an hour. Any train delays, because of a missing itinerary, could cost up to $15,000 per hour!

The Solution

An IBM RS/6000 running AIX was chosen as the fax server hardware platform of choice, because of its robustness and capability of supporting multiple channels, while continuing to queue new jobs and perform administrative functions.

Total compatibility with the chosen operating system was among the chief requirements. Since FaxTrek is natively written for AIX and utilizes Dialogic's GammaLink dedicated fax cards, integration was both swift and smooth. High-volume fax is best done with high-end fax boards and software; they go through extensive testing to ensure compatibility with every other fax device.

Three IBM RS/6000 F30s were utilized, each connected to two (T-1) digital links. Using four Dialogic CP12-SC GammaLink cards and two digital interface boards per machine (see Figure 2), each F30 was equipped to drive 48 fax lines.

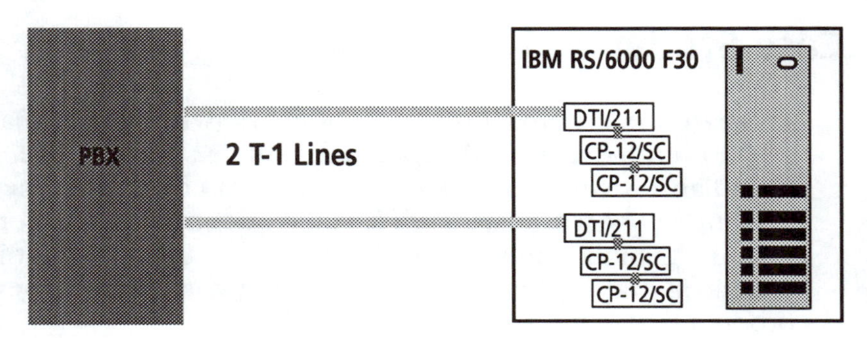

Figure 2

This particular installation, to meet the volume of faxed itineraries, required only 96 lines, so why were three machines implemented? In response to its client's concern for protection against system failure, a third server was employed as a backup server, which was configured with a real-time redundant fax queue.

In addition to the standard FaxTrek features, a comprehensive customized Application Programming Interface (API) was provided to comply with the needs of the project. This provided the creation of a tool-kit that permitted the establishment of a connection with the mainframe host, the redundancy system, and the scripts that are called during the fax process.

Lightning FAX for AIX

Lightning FAX for AIX was developed as a Window's clients in enterprise fax configurations, which employs a modular, true Client/Server architecture through the TCP/IP protocol. Lightning FAX for AIX features clients for Windows NT, Windows 3.1, Windows 95 and JAVA, and offers a Fax Integration Toolkit for easy integration of fax services with legacy systems. (Lightning FAX Server is also available on the Windows NT platform).

The user-friendly interface of Lightning FAX emulates Windows Explorer. If you are familiar with file management in Windows Explorer, managing faxes will be simple.

WAN/Internet

Lightning FAX has retained the mission-critical reliability of FaxTrek, while adding even more scalability. In addition, Lightning FAX is easy to implement and to manage in a LAN/WAN or Internet environment.

Interstar's latest servers and modules use TCP/IP to communicate between applications and workstations. Since TCP/IP is both used with LAN/WAN and the Internet, this means no additional protocols to install and configure. The servers work with Simple Mail Transport Protocol, enabling organizations to integrate email and Fax services.

Adding a server to your existing network is simple: just provide a valid IP address to the server and configure the clients on the workstations and your fax system is operational. This process is almost as simple as adding a network printer.

The modular architecture of Lightning FAX means that Server, Driver, and Clients can be installed on any machine on a network. Network administrators have flexibility to customize their installation for maximum efficiency.

Lightning FAX Server manages up to 15 different Resources, corresponding to fax board drivers, which in turn can operate up to 48 lines of fax each. Each Resource can be configured separately. Each fax Client can be given access to a particular Resource or Resources.

Interstar has pushed this notion to its logical conclusion with the Least Call Routing (LCR) module. Thus transmissions can be automatically forwarded, through a private WAN or the Internet, to a Server in a different location, to minimize long distance charges.

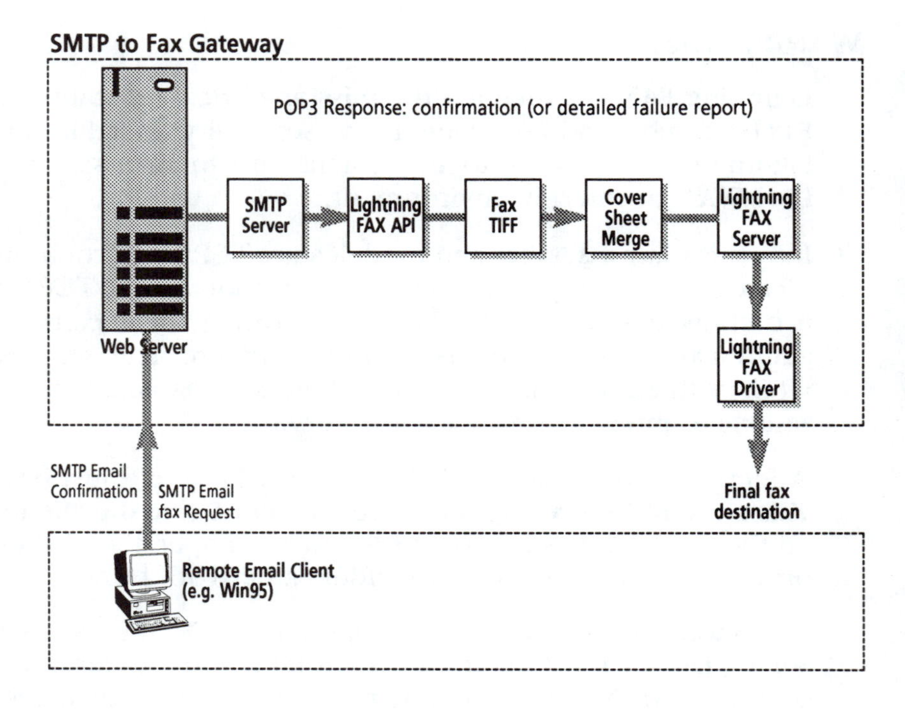

Figure 3

The use of the Internet is not limited to transmission. Interstar's products are customizable via a fax Integration Toolkit. Figure 3 represents a sample configuration that could be used by an ISP to provide fax services over the Internet.

Developing CBF Solutions

Call Center

This module combines a Call Center interface with various Optical Character Recognition (OCR) and imaging applications, working together with the Lightning FAX Server.

The Call Center retrieves faxed documents, such as contracts and applications, from the Lightning FAX Server, applies an OCR process to the form, provides an interface for visually inspecting and editing the resulting data, and sends this data, on confirmation, in a file to other applications on the network. Developed for a major North American telecom company, the Call Center module not only greatly reduces the drudge work of repetitive data entry, but also offers tools for reducing human error in the cycle of forms processing.

Least-Cost Routing

LCR is based on the premise that corporations that fax documents across regions and countries could benefit from the use of the Internet, or their own private WAN, to save transmission costs.

Let's say, for example, that your company has an office in New York and another one in Tokyo and that you regularly need to send faxes to Hong Kong. Normally your long distance charge would be for a call from New York to Hong Kong.

If Lightning FAX is installed at both locations, your fax to Hong Kong would first be routed—through the Internet to your Tokyo Server, then would be faxed through telecom telephone lines to Hong Kong, a much shorter distance, resulting in a long distance savings.

For more information visit: www.faxserver.com

Chapter 30

Production Fax

with CommercePath

The Fax-Enabled Enterprise

Efficiently leveraging business-to-business communications throughout the enterprise can mean hundreds of thousands of dollars in annual savings for a large company. Enabling existing applications to communicate with customers, suppliers, and partners using the document-delivery technology they prefer-phone, email, Internet, EDI, fax, and/or surface mail-can now play a vital strategic role, and can quickly yield impressive returns for an organization.

Fax has emerged as just such an enabler due to its ability to reach a wide audience and because it can be easily integrated with other technologies in the document communication mix. In a recent survey among Fortune 500 companies, for example, fax was the preferred method of delivering urgent documents by 34 percent of those surveyed-outpolling courier services, email, phone, voice-mail, and surface mail. Companies that do a lot of faxing, particularly from high-volume production applications in the back office, have learned that leveraging the power of new fax technologies can mean an ROI measured in weeks, as opposed to the months or years associated with previous methods of conducting business-to-business communications.

To take advantage of this, a change in mindset is occurring. Fax activity is shifting from the fax machine to strategic enterprise-wide back-office electronic commerce communication methodologies. Fax-enabling the enterprise means integrating fax with mainframe, client/server, and LAN computing environments and communicating directly from applications in any of these environments, not just the LAN. It means leveraging the multiple features and benefits available through new production fax integration technologies, including support for high-volume faxing with forms overlay capabilities, EDI-to-fax conversion and delivery, fax functionality from the LAN, the union of fax and interactive voice response (IVR) technology, and delivery of fax over the Internet as a low-cost transport mechanism.

By integrating various production fax technologies on a scalable NT platform, companies can also add new communication functionality as the need arises, further leveraging ROI. And, because new fax technology is integrated with single or multiple servers, it provides new accessibility and value to a wide variety of users throughout the enterprise. From purchasing to marketing, customer support to accounting, fax is emerging as a vital integrated link in the back-office, business-to-business communication chain.

High-Volume Communications with Fast Turnaround

For large companies, the strategic advantage of using production fax technology lies in the ability to automate existing high-volume, mission-critical applications. Each month, for example, Mitsubishi Credit receives and processes 20,000 loan applications from 500 auto dealers nationwide.

With a production fax solution integrated with the consumer lending solution on its IBM mainframe, Mitsubishi dramatically reduced the time it takes to respond to loan requests. As a result, it has garnered a larger share of the consumer lending market. Mitsubishi's production fax server also allows the company to save on printing and mail costs because it is used for automatically faxing statistical reports in batch-mode directly from its IBM mainframe to dealers at night, when phone rates are lower.

Carnival Cruise Lines uses production fax technology to streamline high-volume operations. The company used to spend $15,000 to mail confirmation details and itineraries to thousands of travel agents worldwide. Now, with a production fax solution integrated with applications on its Unisys 2200/900 enterprise server, a Unisys 6000/75 UNIX server, and networked PCs, Carnival distributes that same information for less than $3,000. The company also uses its fax server to distribute 9,000 purchase orders monthly, and by integrating the solution with an internal email system, expedites the processing of change requests.

Conversion Capabilities Overcome Communication Incompatibility

Another common use of emerging production fax technology is the conversion and delivery of EDI documents. Companies that rely heavily on EDI, but conduct business with some non-EDI-capable trading partners have long sought the ability to simply deliver EDI documents as faxes. Converting EDI documents directly to fax for delivery to non-EDI-capable trading partners helps companies reach all their trading partners with equal efficiency and eliminates the expense of maintaining separate systems for both fax and EDI communication.

One large federal agency used to mail 2,500 purchase orders daily to non-EDI-capable suppliers. But by integrating a fax solution with its Unisys mainframe-based purchasing application, the agency now faxes the same letter-quality purchase orders for one-tenth the cost. The result is a $150,000 annual savings. The agency also uses a related solution on the same server to provide fax-on-demand services, enabling customers to use their touch-tone phone to have documents automatically faxed back to their fax machine, avoiding the cost of employing additional customer-service representatives.

Leveraging Fax, Email, and The LAN Increases Employee Productivity

Another strategic implementation of fax technology in the enterprise is via the LAN and on desktops. New fax technology enables users to fax directly from existing desktop applications such as Microsoft Word, Microsoft Excel, Microsoft Exchange, Lotus Notes, and SMTP programs, eliminating the time-consuming process of manually faxing documents. In addition to traditional PC-based fax software, these technologies support high-volume, individualized faxing, inbound and outbound fax management, and automatic routing of inbound faxes directly to the desktop. At a large HMO, for example, a solution implemented in a Windows NT Server-based LAN environment enables physicians to write rush prescriptions in email, but deliver them to pharmacies as fax documents using the same email interface they are used to.

Customized 24-Hour Customer Service with FAX and IVR

New fax technology also uses integrated IVR to provide the rapid fax-back services which are so vital to customer-support organizations. In a typical fax/IVR solution, customers use a touch-tone telephone to interact with a series of automated voice prompts to order up-to-the-minute account statements, ad hoc reports, bank rate sheets, updated price lists, and other documents. The solution automatically and immediately delivers the selected documents to the user's fax machine.

How the Technology Works

Host-based, production faxing works by essentially transforming every fax machine in the world into a remote printer linked directly to a company's legacy, client/server, and LAN applications. A set of fax-command-language (FCL) APIs is embedded into the print applications on the host with instructions for formatting and delivering documents received in the print file. The solution emulates printers in various environments and uses the FCL commands to pull up the necessary background forms residing on the server, places graphics and signatures on them, overlays the data to build each unique document and delivers them to any fax machine automatically. By emulating native printers and other devices, the solution can be integrated with applications on any host system.

The ability to convert EDI documents to fax for immediate delivery is a solution that is easily integrated with a company's existing EDI translator software. Such software accepts, as input, standard ANSI X12-formatted output from any EDI application. After confirming delivery of the fax, the software automatically generates a functional acknowledgment and delivers it to the EDI translator so that all EDI-document tracking takes place within their existing EDI application.

In a LAN environment, employees deliver fax documents directly from their desktop through integration with their existing desktop applications and email package. Desktop users send and receive faxes just as they would email, and can respond, print, store, or forward, the fax.

Fax on demand and interactive voice response support is implemented through a fax server and software that interact with host-, server-, or desktop-based applications to build and fax documents back to users. Broadcast-fax technology retrieves and faxes static documents, while IVR systems query host-based databases to extract and format data on the fly. Such "live data" dynamic IVR systems complement production fax applications by enabling recipients to access frequently updated documents as needed and immediately receive the information via fax.

Another emerging fax technology is support for Internet transport. In one approach, documents are converted into MIME format and delivered to an Internet address over standard phone lines. Some systems deliver faxes over the Internet as email instead of image documents by sending GIF files in MIME "envelopes," with MIME viewers displaying fully formatted documents for recipients.

What to Look For

Companies in search of a strategic approach to integrated back-office communication should consider a number of factors in addition to cost benefits when selecting fax technology. One factor is the architecture of a solution as it relates to the environment in which it will be deployed. Ideally, a solution for an IBM MVS environment, for instance, should be able to emulate 3270 and 3770 devices and TCP/IP, and should support AFP faxing to provide users production-fax functionality with the same control over formatting mixed text-and-image files that is available when printing AFP documents. Solutions designed for a Windows NT Server should run as a true Windows NT Service, and employ various links to popular databases.

Additional considerations depend on the environment. For instance, companies using line-of-business applications from vendors such as Oracle, SAP, and/or Sterling Commerce should seek a solution that integrates well with existing applications. They also should look for support for industry standards in terms of fax cards, graphic formats, network connectivity and topology, routing options, print types, and EDI formats.

Functionality is key, of course, and different solutions provide different levels of functionality in selected areas. Purchasers should consider the kind of work they'll be doing with the solution-now and in the future-to best evaluate support for intelligent inbound and outbound email interfaces, fax-on-demand, IVR, Internet capabilities, and EDI conversion. Productivity features that enhance functionality include automatic document batching, which combines multiple documents bound for a single destination into a single fax session to minimize call set-up time, and host-notification, including host-based capturing and communication of performance statistics for automatically updating applications and databases on the host.

Following the lead of enterprises moving to client/server environments, solutions based on emerging production fax technology include a number of sophisticated administrative capabilities. Such capabilities enable users and administrators to fax both static and dynamic documents directly from desktop or production applications on the LAN, to route inbound faxes to individual desktops without manual intervention, and to execute large broadcast-fax runs from the desktop.

Other administrative capabilities include remote diagnostics and alerts to provide feedback such as tracking statistics between the fax software and the host, and desktop capabilities for viewing, enhancing, printing, routing, resending, forwarding, and deleting inbound faxes. Also beneficial are centralized capabilities for monitoring, resending, advanced sorting and storing, building large phone books, faxing large broadcasts, querying lists, sorting transmission logs without Windows memory constraints, alerting an administrator when a fax is processed, and tracking fax traffic from individual PCs and from the fax server.

Performance is fundamental to the ability of any fax solution to fill its strategic promise. Hence, companies considering such a solution should look at support for forms caching that limits form-and-data merge time to an average of two seconds per page. For fast transmission, a solution will ideally use G/3 and G/4 compression methods, MR, MMR, load balancing, and a driver-level software/hardware interface.

Fax as a Strategic Advantage

With the vast array of capabilities offered by today's emerging fax technologies, large companies are finally in a position to deploy electronic-commerce communications strategically, just as they have long done with EDI and, more recently, with email and the Internet. Whether they're fueling production faxing, forging a link between EDI- and non-EDI-enabled companies, supporting LAN-based fax routing services, founding a customer-friendly IVR/fax-back system, or employing the Internet for fax transport, these technologies create robust solutions for mission-critical, business-to-business communications. A company deploying such technologies-becoming "fax enabled"-learns the strategic value of fully automated back-office electronic commerce: the ability to do business with anyone, anywhere, without the traditional limitations of information format or document type.

Companies involved in providing back-office solutions for production faxing, such as CommercePath, inc. in Portland, OR, the leader in this market, provide a scalable set of solutions on the Windows NT Server platform along with the scalability to grow with a customer from entry-level to large load-balanced systems and solutions that can be customized with a varied set of modules to meet the unique needs of numerous environments.

Since 1990, servers running CommercePath products have been installed worldwide, supporting applications such as invoicing, purchasing, loan processing, and order confirmation. CommercePath customers include Bank of America, Carnival Cruise Lines, Federal Express, Ford, Harley-Davidson, PepsiCo, Reynolds Metals, and Toro.

For more information visit: www.commercepath.com

Chapter 31

Fax Broadcasting

with Copia International

Introduction

Copia is a of private and Commercial Broadcast Fax software. There are service Bureau's that are larger, but they have internally-developed Software. Copia provides fax software solutions for customers needing 2 to 99999 phone lines.

This chapter will cover the design issues for selecting and running a fax broadcast system. After reading this chapter you will better understand the problems and Copia's approach to solving them.

Copia's History in Fax

Copia was the First Company to provide a Fax-On-Demand system as a software product. Copia released FaxFacts in

November 1989 at COMDEX in Las Vegas NV. Copia FaxFacts systems have been in continuous operation since 1990. Copia was the inventor of the same call Fax-On-Demand system, and was granted a US patent for this invention. In 1990 the fax and voice boards handled only a single line and we had to develop a system that could share the load across multiple CPU's. The early design decision enabled the fax load to be shared.This early design has paid off for our fax broadcast systems, by allowing very large systems to be installed.

Some of the questions you need to ask yourself in planning your broadcast system are:

Issues When Planning Broadcasting Systems

Is it scalable?

Can additional CPUs be added to deliver faxes and can additional CPUs be added to help with the rasterization process, in features such as mailmerge?

As we continued to work on the Fax-On-Demand system, we decided to hand-off "faxback" requests into a general queue. This queue was then processed by all the CPUs acting as "fax servers." The queue is NOT a file, but a directory that we call "TOSEND." One of the most important design specifications of the FaxFacts system was to be an "open" system. When you create a queue that is a file with records in the queue file, you now have to provide an API to allow outside programs to send faxes. Each programming language needs its own API.

Copia supports the entire Dialogic/Gammafax productline along with Brooktrout, Puredata, and others. We decided to store the information needed to send fax in a normal ASCII file. Using a file to store all the information that would normally go into a "queue record" has many pluses and some negatives. With an ASCII file per queue request, we have the ability to write a request to send a fax from any programming languages. The essence of the FaxFacts API is the following information written into a file:

$fax_phone 630-778-8848
$fax_filename mytiff.tif

The above file is written to the tosend folder and given a unique file name. The FaxFacts Engine controls the broadcast process in each CPU that has fax boards in that CPU. The FaxFacts Engine looks for files that are not claimed in the TOSEND folder. When the Engine finds a file in the TOSEND folder that is unclaimed it starts to send the file named in the file to the phone number specified. When the fax is sent the file is then moved from the TOSEND folder and placed into the SENT or FAIL folders. This has the advantage of the TOSEND folder constantly being emptied by the Engine. The The file stores additional information such as the failure code or the CSID of the receiving fax machine. This additional information can grow without the limits of field size that a queue record would impose.

The design of our file-based queue also removed the need to have to pack, compress, or run maintenance programs on the faxing system.

The system duty cycle should be 7 by 24

This means that the system should not have files that grow and then need to have programs run to "clean-up," "remove gas," "pack deleted," or "make faster." Some systems can not be run for days, week, or years with out having a person adjust something. Copia has designed FaxFacts to run for years without any "operator intervention."

Why Three kinds of fax broadcasting?

With FaxFacts, a broadcast can be one of three (3) ways. When you send the same image to a list or group of people, we call that a fax blast or simple fax broadcast. If you want to increase the reader attention to the fax, you can do a Graphical Coversheet broadcast. With the Graphical

Coversheet broadcast you can place database field anywhere on an image template (we call this a watermark). The result is a fax that is sent with the basic message and the variable information from your database overlaid onto the watermark. Each field can be positioned size, font and angle can all be setup using the visual fax viewer to design the final fax. If you need even more power you can do a Mailmerge to fax using the FaxFacts FFMERGE system. The Mailmerge system is unique to Copia in the way that we have approached the problem, more later about FFMERGE.

Fax Blasting

While FaxFacts is a great Fax Blaster for people that have just two phone lines, we are much better than the others when it comes to larger numbers of faxes to send. Earlier I said that FaxFacts is the "no limits" faxing systems. What that means is that when you want to Blast Fax to 1,000 10,000 or virtually unlimited number of recipients.

When Blasting is there only one copy of the fax?

Many systems have to have a separate image for each person that is going to receive a fax. This causes the system to have to create, move and/or use more disk space for each fax receiver. With FaxFacts each receiver gets a small file that contains the name of the fax image to be sent. This saves time and space on the disk. Using the FaxFacts Fax Blaster we can launch 100-150 faxes per second. This means that you can launch 9000 fax blasting requests per minute! This does not mean that you are sending 9000 faxes per minute, but that you have queued 9000 fax requests. The launch speed only becomes a problem if you can not launch faxes as fast as you can send them. How many lines would I need to have sending faxes to keep up with 9000 faxes per min? If you figure 45 to 60 fax pages per hour. You would have to have about 9000 phone lines to keep up with the launcher. With "no limits" faxing, when you launch a blast fax you are not limited by the file system as to how many faxes can be launched. When you launch a blast fax under FaxFacts the only limit that can be reached is when the "file server" is out of disk space.

Graphical Coversheet Broadcast

The next type of broadcast faxing is what we at Copia call Graphical Coversheet broadcast. This type of blast fax is designed to increase the readership of direct mail via fax. I think that most of you have seen a direct mailing that had your name in the body of the sales pitch. This same effect can be done with Copia's Graphical Coversheet blast fax. We have had good luck placing the name of the person we wish to read the fax on a 25-degree angle in a script font. It says Steve, I thought you would be interested in this info. This makes it look like some friend of mine saw the information, wrote a note on it and faxed it to me. The chances of me reading the fax are much greater when my name is written on the fax.

Mailmerge to Fax

Copia's real strength in fax blasting is our FFMERGE product. We were asked if we could do mailmerge to fax and in the beginning, we couldn't. The problem with mailmerge to fax is how do we get the information from the wordprocessor when one fax stops and the next one starts, and how can we pass the information of the fax number and other information. We were aware of other solutions that use DDE(Dynamic Data Exchange), vendor specific API calls, and embedded keywords.

We decided to create our own TrueType font. But instead of catching the font on the way into the windows conversion to graphical raster lines, we would look at the graphical raster lines as they came out of the rasterization process.

We designed our font to have a line at the top of the character and a line at the bottom of the character. Just under the top bar we put a binary code that we can decode into the actual character. We took advantage of the following information. A 12-point font is the same as 10 characters per inch. At 200 pixels per inch for fax images that gives us exactly 20 pixels or dots across each character. We make the overbar 18 dots across the top and bottom of each character. The overbar allows us to "see" our font at the start of the page.When we "see" our font we are able to read the next scan line and decode it directly into the information needed.This line we call the "firing line" and it contains the fax number and any other information that we wish to pass from the printing system to the faxing system. By only having to look at the top information to find the "firing line" the rest of the printing process goes along at full speed.The FFMERGE printing speed to create the ready to fax tiff image is 30 to 60 pages per minute. (this differs from the 100-150 mentioned earlier) Let's see how well our invention of FFMERGE meets our design goals: Is it easy? Yes all you have to be able to do is get your mailmerge running to paper, then place the fax phone number field at the top of your mailmerge form letter. Set the font to the FFMERGE font that we supply and set the size of the font to 12 point.You are now down with the setup of FFMERGE.

Do we have to program? No because we are looking at the print stream as a print driver, we do not know or care what windows application that is being used.We now do not have to purchase and test each and every word processor on the market. The customer is not bothered by needing to get operational the "macro package" for each new version of the OS or word processor.

What databases do you support? This is a common question that we get with our mailmerge solution. Because of our design getting our information off the printed image, we support any database that the word processor you are using will support.Again, if you can print it, you can fax it.

Do you support the latest version of Word (Office 97), or the release that just came out. Yes, If the system prints to Windows and you can control the font selection, you can use FFMERGE. (Copia was issued patent number 5,715,069 on Feb 3, 1998, for the FFMERGE product)

*** Sure we can do mailmerge, NOT ***

We have had a number of people come to us saying that they

purchased a product that could do mailmerge. BUT whenever an application, for example MS-Word, was upgraded, this seemed to cause problems for their mailmerge software. Also some implementations tend to slow down as the size of the mailmerge grows. I have had reports that the time for each page at the start is very quick, but after 1000 records the system was so slow that it could not keep up with the number of outbound lines sending faxes. Just because the vendor says that they can do mailmerge to fax, be sure to ask if it works with current tools and the latest OS.Also what are the defacto limits to the size of jobs that the systems "likes".

After you are amazed at the ease of use of the FFMERGE solution, you will see that it is "pop-up" free and MUCH faster than any of the other systems that are available.

FFMERGE extra credit

After you understand how the FFMERGE print driver works with the True-Type FFMERGE font, what about all the other things that you print? Why would you ever mail something that you print from you computer? With the FFMERGE tool, and invoice, statement, or purchase order can be printed directly to the fax machine of the person that is intended to receive the information. To me it makes no sense to print a piece of paper, fold it, address an envelope, place postage on the envelope and leave in in the hands of the US post office. I don't know about you but I am from Chicago, we burn mail in Chicago! Using FaxFacts to deliver items normally mailed, gives you access to the receivers fax printer on a 7 by 24 basis. The US mail people have defined "on time" delivery of mail from one suburb to another to be two days.

We are working with people that are pulling up their payment times by using fax and also lowering their costs for delivering paper to other businesses. Helpful tips and hints: How many phone lines will I need We use as a rule of thumb 45 pages per phone line per hour, for a one page normal mode fax. Even though the fax may only take a minute or less to send, 45 pages per hour is closer to the actual time per line because of busy time, dial and ring time, adding to the time to send a fax. If you are going to be sending more than one page as part of the same fax you will see the number approach 60 pages per hour per line. Or if you are sending High Res faxes to get a better look, the number will drop to closer to 30 pages per hour per line.

How much disk space will I need?

The more the better. If you are using either Graphical Coversheet or mailmerge to fax, you will be creating two files per fax receiver. The TIFF file will be from 30,000 to 100,000 bytes and the control file will be about 300 bytes. If you have a large disk and the file system is FAT, not Novell, or NTFS, you will use more disk space than you expect. With a FAT(File Allocation Table) file system which was the original MS-DOS file system, even small files may take 32,000 bytes. The problems is when you write a 300 byte file and a 40,000 byte TIFF file, the system will allocate 3 blocks of 32,000 bytes. While the DIR command shows 1000 300 byte files, you will see the amount of disk space go down by 1000 times 32,767 bytes. On a Novell system the allocation block size is 4096 bytes per block if you are not compressing the file system. This is NOT due to FaxFacts except that this is one of the issues when you have lots of small files. We feel that the benefits of load sharing and speed overshadow the wasted space due to file allocation issues. One of the best files systems for the FaxFacts system is the NTFS system that has a typical allocation size of 512 bytes. The 512 bytes is close to the actual size of the control files that we use.

Smart Retries

As we at Copia have been working with different fax boards and customers, we have learned about how to get faxes through if possible, AND not try forever to deliver faxes to number that will never work. We started with specifying a number of retries and a time delay between tries. This is what many of the systems offer. The first problem that we saw that was not addressed by this design, was "self induced busy signals."

Does the system detect and prevent its own busy signals?

A "Self induced busy" is when you have more than one person in or broadcast list at the same phone number. We had more than one phone line that was trying to send to the same fax number. When I looked into how other systems solved this problem, I found that they only were able to increase the number of retries to fix this problem. What we have done is to prevent the problem in the first place. As each phone line selects a file to send a fax, it first checks to see if the phone number is in use by another phone line. If the phone line is being used by another fax line, then the file is skipped, and the retry count is not incremented.

Can the system restart in the middle of the fax?

After fixing the busy problem we have improved the sending of faxes with what we call "Smart Retries" The FaxFacts "Smart Retry" feature provides enhanced control over retrying failed outbound faxes. With this feature, you can specify that only the un-sent pages are retried if a transmission fails in the middle of a sequence of documents, and you can specify a different pattern of retries for different classes of failure. A special cover sheet can also be specified for retry attempts. The entire system is table driven, so that you can modify the actions that the FaxFacts system takes for each and every failure status that is returned from the fax board system. You can choose to retry specific error codes or fail them and not retry more than once. One use of this feature is to have the fax be retried, if it is busy or ring no answer, after an 8 or 12 hour delay. This will allow faxes to be delivered to those fax machines that are out of paper or turned off in the evening.

Is the system automated?

The Copia FaxFacts system as always been focused on being able to have the system "run itself." One of the great features of Windows, is that it is visual, and graphical. The graphical nature of faxing is great for Windows and the mouse is great for working with fax images. The problem with the Windows/mouse interface is that it requires a human to move the mouse. In addition to the visual tools to launch and manage fax broadcasts, FaxFacts also has automation versions of the tools. The automation tools allow fax broadcasts to be launched from IVR, email, receiving a fax, and time of day. Many of our service bureau systems have completely automated the launching of customer broadcasts. The customer first uploads the list and image via fax, BBS, or email/web. Then customer then receives a proof fax and time to call up an IVR port and kill the job if the proof is not of the quality needed. The FaxFacts system options allow IVR input prior to receiving a fax. The IVR information collected specifies the list, priority, and time to send the broadcast. The customer can also call into the system and get the progress of the job either via IVR voice playback, or a faxed status sheet.

Do Not Send (DNS) Block fax numbers

The system needs to have a central ability to block fax numbers. The FaxFacts system has a, high speed, lookup table that has the list of all phone numbers that are to be blocked. The FaxFacts system has two levels of fax number blocking. There is a central DNS file and then when you launch a blast fax you can specify list or customer/group blocking lists. Our service bureau customers have central DNS files and customer specific DNS files. This is a big issue due to the federal law that can cost you $500.00 for each fax that you send to a person that does not want to receive faxes from you or your company.

Fax Application Programming Interface API

The fax broadcasting software that you buy MUST have some kind of API (Application Programmer Interface). Having the API will make sure that the system will be able to grow and meet your future needs for faxing. Some systems try to use a "standard" interface, such as CAS, or the GammaLink API. The FaxFacts API is a higher level API that is the same API for CAS systems, GammaLink, Dialogic, and

Brooktrout.

The highest level API from FaxFacts is the FFMERGE system. All that is needed for a programmer to send a fax is to be able to print the fax number at the top of the printed document. The programmer can also specify attachments and other information on the FFMERGE "firing line". Using the FFMERGE print driver and font to control faxing is very powerful in the hands of a programmer. The benefit of this high level API is that it does not limit the choice of tools that can be used to do the job.

The next level of API gives the programmer all of the control and feedback that is needed. The FaxFacts API is easy to use, has many defaults, and has all the power that programmers ask for. Now for an example:

In the file PO980315.fs

```
$fax_filename \\faxserver\copia\temp\po980315.TIF
$fax_phone 6307788848
$fax_status1 2 ;ready to fax
$fax_origin user_request
$fax_user \\faxserver\copia\fax.usr ; group file with defaults
```

When the programmer writes the above file from their application program, they can monitor the progress of the fax. Once the file is written to the TOSEND folder, the sender will know when the fax has been sent when the file is removed from the TOSEND folder and moved to the SENT or FAILED folder. If you do not want to monitor the fax directly, you can specify a "post-process" program to be run by FaxFacts when the fax is "complete". By "complete" that means that the fax has been sent, or that it has completed all the retries specified and is being move to the FAIL folder. FaxFacts customers use the post-process to send email back to the sender. The post-process is used to update databases or to do what ever work needs to be done when the fax is complete.

If more control is needed, additional API commands are

available.

$fax_cover — sets the coversheet to use if needed.
$fax_header — sets the line at the top of each fax.
$fax_pre-process — Specify the routine to run before faxing.
$fax_post-process — Specify the routine to run after faxing.
$fax_retry — Set override retry settings.
$fax_send_time — Set the send time in the future.
$fax_send_date — Set the send date for the fax.
$fax_send_line — Specify the line or line group to use to send this fax.
$fax_sender — The name of the sender of the fax.
$fax_receiver — The name of the fax receiver.
$retry_cover — The coversheet for restarted faxes.
$var_def — Define a user variable for the cover sheet.
$voice_phone — Call out for IVR or pager.
A more complex example:
$fax_filename t:\faxfacts\image\00001154.1
$fax_cover t:\faxfacts\copia.cvr
$fax_header "To @ROUTETO From: @SENDER"
$fax_sender "Dorothy Gaden"
$fax_receiver "Ann Other"
$fax_send_time 17:00:00
$fax_phone 3105553218
$fax_status1 2
$fax_post-process "exchange_send" internal
$fax_origin user_request
$fax_user t:\faxfacts\fax.usr
$var_def ToCompany "Other Company Inc."
$var_def data1 "recid1000"
$var_def user "email address"

Smart fax boards vs. low-end fax boards.

Copia's FaxFacts system supports the high level fax boards from Brooktrout, GammaLink, Puredata, and Dialogic. The high level fax boards have a computer for each telephone line. The computer on each phone line helps keep the remote fax machine happy that the sender has not stopped sending when the main CPU may be busy. With the low end modems, when the main CPU becomes distracted, the remote fax machine may think that the sending fax has stopped sending and drop the line. This causes a higher failure rate and more phone costs. The larger issue is that if you have problems sending to a specific fax machine and the low end fax board fails every time, who do you call to fix the problem? The problem is that on the low end modems there is very little that the supplier can do to fix fax problems. With GammaLink and Brooktrout they have downloaded firmware that they can and will fix. Now that all of the fax boards send the same faxable TIFF files, you now can use the common API from FaxFacts and even mix fax boards from different manufactures in the same system.

Fax-On-Demand

FaxFacts also has as an option Fax-On-Demand. Unlike other systems that have added FOD as a late add-on, FaxFacts started as a FOD system. Many of the high end Dialogic and Brooktrout fax boards can do both voice and fax on each phone line. FaxFacts gives you the option of having Fax-On-Demand, IVR, Voice broadcast, Fax Polling, Internet faxing, dialer faxing, web page faxing, faxing from your web page, and faxmail. FaxFacts has more features in the FOD area than any other FOD vendor.

For more information, visit: www.copia.com

Chapter 32

Enterprise Fax-On-Demand

with Castelle/Ibex Factsline

Introduction

The Ibex Division of Castelle builds fax-on-demand and high capacity fax systems and has been in the market since 1989. They have been extremely successful in the high technology market so if you have ever requested an automated fax from one of the large software companies, you have probably used an Ibex system. Because of the leadership position and time in the market, Ibex has been able to create and market ancillary fax delivery products such as automated forms processing, Lotus Notes integrations, fax server, high volume fax broadcasting, host interfaces, and other products.

Ibex and its fax software products were acquired by Castelle in 1995. In addition to these enterprise fax products, Castelle also markets a turn-key hardware based fax server, Faxpress, targeted to small and mid-sized companies as well as departmental use. FaxPress is covered in another chapter in this book.

Beyond Fax-on-Demand

Because the World Wide Web and email are now used as extensively as fax for information delivery, Ibex has used its strengths in fax processing and document management to build a "universal information on demand" system. The system of course includes fax-on-demand, but also uses email-on-demand and the Web to deliver documents. This book, being a book on fax, will emphasize the fax-on-demand portion of the product.

One-Call Fax Delivery Mode

The Ibex system supports two types of fax delivery modes: "One-Call" and "Two-Call". One-Call mode means that the caller must call from their fax machine, and the resulting fax transmission is sent back on the same telephone line. Callers can call a Two-Call system from any phone—the system will prompt the caller for a fax telephone number to be used to send the fax. Its called Two-Call (or Call Back on some systems) because the system will deliver the fax with a second call. You may mix One-Call and Two-Call modes on the same system. For example, you may wish to designate that some incoming lines operate in One-Call mode, while others operate in Two-Call mode, or you may even specify that a voice line may operate in One-Call and Two-Call modes depending on the time of day, who the caller is, or some other criterion.

An important aspect of systems set up in One-Call mode is that the system throughput will not be as high as a similar system (same number of voice and fax ports) set up in Two-Call mode. This is because a Two-Call system can have all voice ports and all fax ports busy at the same time. A One-Call system must have fax ports idly standing by whenever a voice call is in process. In fact, most One-Call systems including Ibex's physically connect each voice port to a fax port, thus guaranteeing that there will be a fax port available when the voice port requires one. Other important aspects of the One-Call method include:

- Phone usage costs are lower since the fax transmission takes place on the caller's initial phone call.This needs to be weighed against having to purchase more voice and fax ports.
- Useful for international callers, since they do not have to enter their international phone number as the system would have to dial it from the host country.
- Callers must call from a fax machine in order to receive a fax.
- The phone line is tied up for the duration of the fax transmission. No new phone calls can come in on that voice telephone port.

Two-Call Fax Delivery Mode

The Two-Call method requests a fax telephone number from a caller after a fax has been chosen.The fax transmission is sent immediately if no other faxes are waiting, or it is queued if all available fax boards are busy.With this method, the system will process incoming calls at the same time that it faxes out requested information.Thus it is possible (with a four voice, four fax line system) to have four callers listening to voice menus and making choices while four previous callers are receiving faxes. Other important aspects of Two-Call are:

- More convenient for callers since they do not have to be at a fax machine to use the system.
- Can be used to send faxes to someone other than the caller.
- Costs more for the FOD system owner in that all fax transmissions are separate telephone calls initiated by the FOD system.entered by callers.This is especially true for international callers who must know to enter the country code, city code and phone number to successfully receive a fax.

Although a significant percentage of FOD systems used One-Call FOD during the beginning of the FOD market in the late 80's, most Ibex systems installed today are Two-Call. In fact, one of the largest and oldest FOD applications (the Internal Revenue Service) is still One-Call to Two-Call. However it is still used where cost is important and also for international applications. The remainder of this section concentrates on Two-Call fax delivery.

Sizing a System

The size of a FOD system is measured in ports, and is usually broken down further into voice ports and fax ports. For example, an Ibex 4x4 system contains four voice and four fax ports. An average FOD system has the same number of voice ports and fax ports (average voice call is two minutes and the average number of pages is 2-3, or 2-3 minutes of fax time). It is common, however, to require more fax ports than voice ports if the faxes sent tend to be lengthy and/or you plan to implement fax broadcasting as well. It is less common to have more voice ports than fax ports; this is usually seen in applications where IVR (Interactive Voice Response) features are used which can lengthen the caller's time on the voice port.

The following table sizes a typical FOD system that experiences an average call time of two minutes. To use the table, first look up your average fax time across the top of the table (use one minute for one page, unless you know you are going to be faxing lots of graphics or use high resolution, then use 1.5-2 minutes per page). Then look up the maximum, peak number of document requests you expect in an hour along the left hand side of the table. The table will then provide the number of voice and fax channels required. Sizing a One-Call system is quite different—this table sizes only a Two-Call system.

Fax Time in Minutes

Requests per hour	2	3	4	5	6	7	8	Voice Ports
	NUMBER OF FAX CHANNELS REQUIRED							
5	1	2	3	3	4	5	5	2
10	2	2	3	4	5	5	6	3
15	2	3	3	4	5	6	7	4
20	2	3	4	5	6	7	7	4
25	3	4	4	5	6	7	8	5
30	2	3	4	6	7	8	9	5
35	2	4	5	6	7	8	9	6
40	3	4	5	6	8	9	10	6
45	3	4	5	7	8	9	11	6
50	3	4	6	7	9	10	11	7
55	3	5	6	8	9	11	12	7
60	3	5	6	8	10	11	13	8
65	3	5	7	8	10	12	13	8
70	4	5	7	9	11	12	14	8
75	4	6	7	9	11	13	15	9
80	4	6	8	10	12	14	15	9
85	4	6	8	10	12	14	16	9
90	4	6	8	11	13	15	17	10
95	4	7	9	11	13	15	17	10
100	5	7	9	11	14	16	18	10

Table Courtesy of Ibex Division of Castelle

Assumptions for Two-Call system sizing:
Fax time based on exponential distribution
Five percent of fax requests delayed five minutes
Formula: #channels = (fax time /60) * (36 + requests / hour)
70% of incoming voice calls result in fax request
Incoming voice calls last two minutes (mean duration)
One percent of all callers get a busy
Formula: Erlang loss tables

A Real-World Application: Symantec

Symantec Corp., developers of applications and system software such as Norton Utilities, Q&A and ACT, installed an Ibex system in 1991 after experiencing a telephone support crisis. The company released two upgrades of key products. Call volume jumped from 20,000 phone requests to 50,000 in one month. Callers experienced longer than average waiting periods and the company was loosing calls, due to hang-ups. Simultaneously, customer service representatives were manually faxing dozens of product information sheets per day.

A reseller of Ibex systems, Epigraphx, installed a 24 port system at Symantec in 1991 using Ibex's older DOS based software with Dialogic D/41 four port voice boards and GammaLink XPs one-port fax boards. The system was upgraded and expanded by FaxMax in 1994 using Ibex's Windows based software and GammaLink CP4LSI four port fax boards. The computers are racked AST 386 and 486 PCs on a Novell 10baseT Ethernet network, supported by an uninterruptable power supply and tape backup system. The racked FOD equipment is connected to the company's corporate network using a bridge router. Ibex's Interactive Forms software was also added in order to automate the processing of forms that were faxed into the system. The system is maintained by a project coordinator, and requires approximately two hours per week of attention for installing and testing new files, archiving and backing up, and transaction report generation.

Symantec reports, to date, the system has sent over 750,000 faxes.

Boards and Configurations

Ibex systems use Dialogic voice and fax boards (the fax boards were formally known under the GammaLink brand name until purchased by Dialogic). Dialogic two-port voice and one-port fax boards can be used for small systems, while Dialogic high-density boards be used for larger systems. A system can be set up in a standalone (non-networked) configuration using a 386 or better computer supporting up to 24 ports of voice and fax.

In order to be able to add documents from other computers in real-time or modify applications on-the-fly, the Ibex system must be installed on a network. Novell, Microsoft Windows NT, LAN Manager or other full service networks are recommended. Peer-to-peer networks such as Windows for Workgroups may work for small applications, but are not recommended because of the high network traffic that the voice and fax applications can produce.

Telephone Interface

Typical FOD installations use analog lines either straight from the telephone company (POTS), or from the PBX. If the lines are from the PBX, they need to be analog lines (like those used for a fax machine or modem) and not digital or proprietary lines. Either way you will want your voice lines to roll over to the next line available line (called a hunt group). If you have a large system with more than one computer with voice lines in them, you will want to alternate the roll over lines between the computers to build in redundancy.

Dialogic T1 interface boards (or AcuLab ISDN boards for Europe) are used for large capacity systems. In this case one computer usually supports 24 ports of voice or fax, and multiple networked computer chassis are used to create the system. DID/DNIS, and ANI are supported and can be used to enhance the application.

Since a Two-Call system has separate voice and fax lines, they can be treated completely separately. For example, the incoming voice lines can be routed through your PBX and auto-attendant, and the fax lines go straight out to the central office. Or you can use a T1 card with the voice lines, but use analog lines for the fax ports.

Application Setup

Most IVR and FOD systems use a scripting language to generate applications. Ibex systems break this tradition and uses a data-driven architecture. A Windows front-end provides an easy-to-use interface to modify the application. This design allows the application or document storage to be changed in real-time without taking the system off-line. The record-locking, multi-user aspect of the databases allow for contention and simultaneous changes.

Voice Menu

The core of an Ibex application is the Voice Menu, a voice announcement that allows callers to perform actions by pressing a key on their touch-tone phone. The Ibex configuration software allows you to configure a voice menu within Windows.

The right hand side of the Voice Menu represents the telephone keypad. For example, right now as currently configured with this Voice Menu, if the caller presses key one, they will be sent a fax as configured in Fax Box named "INDEX". If the caller presses key two, they will be sent a fax as configured in Fax Box ADOCNUM@ (this Fax Box actually allows the caller to enter document numbers). If the caller presses key zero on their telephone, they will be transferred to an operator.

To assign an action to a key, you would just click with the mouse pointer on the key you wished to configure. A list of action options will then pop up. The available actions from a Voice Menu are:

- Go to another Voice Menu
- Send a Fax
- Record a message (simple voice mail)
- Message retrieval
- Call transfer
- Schedule actions based on time of day
- Host Interface option

- Interactive voice response option
- Credit card processing option

Once the caller actions are configured for a particular Voice Menu, a voice announcement needs to be recorded which presents the options to the caller. These can be studio recorded and imported, or they can be recorded on the spot during the application creation. The Ibex software supports either recording direct to the Dialogic voice board (call from a telephone), or you may record the announcements using a SoundBlaster compatible sound board. To initiation a recording, you would click on the "Announcement" button in the Voice Menu. This brings up the voice recorder, plus the written script for this announcement.

Once the announcement is recorded, the Voice Menu is ready to go. For more advanced applications, the "Options" menu of the Voice Menu allows access to password options, multiple languages, logic (IF-THEN statements), and other options.

Fax Selection

After the Voice Menu(s) is created, you need to define how callers can request faxes. For a simple application you select how many faxes the caller may receive, the length of the document number and the cover page. For more advanced applications, you have the option to choose from the following:

- Which prompts to play, or to customize prompts
- Limit access to specific document group
- Change header text depending on various factors
- Use different cover page depending on various factors
- Use fax number contained in caller account database
- Use binary file transfer instead of fax transmission
- Change number of fax attempts or retry delay
- Activate credit card processing

Document Management

One of the most time consuming tasks with any FOD system is the addition, deletion and management of fax documents and the index of available documents (although not required, many FOD systems have an index of available documents that is available to callers). Ibex systems provide a lot of value in this area by offering Windows print drivers, document management tools and automated indexing tools.

The following document types are directly supported by Ibex (no manual pre-conversion necessary).

- ASCII Files
- TIFF Group 3 / F (Fax TIFF)
- TIFF Modified Group 3
- TIFF Group 4
- PCX
- FMT (Cover page format file)
- FDL (Fax Description Language Files)
- Windows-based documents such as Word
- Web based HTML documents

The images are managed with the following screen from the Ibex configurator. The Description and Category fields, for example, are used to create the automatic index of available documents. The Description can also appear on the cover page, if you enable that feature.

If the "Accounting" button is depressed, then you may assign an owner (supplier) of the document, a meter value which will automatically count and limit the number of access, and the value of the document for credit card charging or for reporting purposes.

If the "Schedule" button is depressed, then you may schedule the new document (or the document modification) to be activated at a later date. Or you can schedule a document to be expired at a certain date.

Adding Documents to an Ibex System

Faxing-in

If the document is not available in electronic format, the easiest method is to fax the document into the system. This can be done remotely from a different site (i.e. New York to San Francisco) or it can be done locally with the keyboard. The image quality will be the same as if you manually faxed the document to the caller -i.e. not the best quality that can be obtained.

Scanning

Scanning is similar to faxing in documents with a slight increase in the quality of documents. You would expect a significant quality increase when scanning, and you can get great results with the right techniques, but with most scanning software doing a Aone-pass@ approach you will get about the quality of a faxed document.

Importing and Conversions

The Ibex Viewer program that views fax images also converts images from one format to another. Most programs in Windows, DOS or the Macintosh can produce a image in .PCX, TIFF, or .BMP format, which the Ibex supports. There are also third party conversion programs, like Hijaak, that support more file formats.

Windows Print Driver

The Ibex Print Driver will create fax documents from any Windows program such as Word for Windows, WordPerfect or PageMaker. You would call up the document using your application, choose the Ibex Print Driver as your active printer, and print. If the workstation is on the same network as the Ibex System, then the print driver will communicate with the system and add the document to the FOD library in one step. In the process you can also specify the document management options (such as scheduling or pricing) as described above. If you are not on a network, then you may create a TIFF image file to be copied or carried to the Ibex System.

Index of Available Documents

The Ibex system can automatically create an index of documents for callers. You can place all documents in one index, or create separate indexes using different groups of documents. The document description, document number, category, number of pages, and the date the document was last updated can be included in the index.

Advanced Topics

Most FOD systems require only the simplest of voice menus (indeed, the simpler the better), static non-changing documents and simple fax queuing and management. However as the industry grows and systems grow in size and importance, new features and capabilities are required.

Fax Description Language

High impact fax documents typically require a mix of constantly changing textual data and graphic images. For example, you may wish to merge a person's name onto a fax-on-demand or fax broadcast cover page using nice fonts and graphics. Or you may wish to create invoices or other forms from dynamic data.

Intelligent fax boards such as the GammaLink, Brooktrout and Intel fax boards will convert text to fax "on-the-fly" by using the on-board processor of the fax board. However, they convert the text to a fixed-size courier type font that is rather bland. Moreover, they cannot mix text and graphics in one location. They can send a horizontal band of graphics then switch to text for a portion of the page, but the result is a nice graphic document with glaring non-graphic text above or below it. Thus there is the "standard" fax-on-demand cover page look—a nice graphic header that drops into straight text.

Ibex's Fax Description Language (FDL) produces full-graphic cover pages and fax documents with unlimited combinations of font types, sizes and graphics. FDL allows you to mix text and graphic elements on a fax page by allowing you to describe the font, size, color (black, white or gray) and placement for the text elements and the placement of graphic elements. The text can be static text, or you may specify variables that may be fax-on-demand caller information, broadcast information for fax-merge features, or data extracted from a database. You can also perform basic drawing functions like creating lines, boxes and circles on the page using various pens and brushes. For example, you may wish to draw a black box on the page with white italicized text word-wrapped inside the box.

Host Interfaces

Host Interface capability allows the caller to obtain information from another personal computer, a mini-computer, or a mainframe. The information is usually based on DTMF digits that the caller entered. These entries may be as simple as an account number, or as complicated as a series of variable length numbers the caller enters in response to prompts. The Ibex Host Interface option is flexible enough to be used with an extremely wide variety of information sources and information formats.

Fax Queuing

As systems grow in size, shaving a few seconds off each transmission and managing the fax resources efficiently can produce significant savings. The Ibex Transaction Manager (ITM) provides these features, and is designed for larger systems that use more than one computer to create a virtual FOD system. For example, if a fax transmission fails after two pages were sent, ITM will only send the remaining pages, not those already sent. ITM also implements an intelligent retry strategy by analyzing the information returned by the fax board, and will not retry the transmission if it determines the receiving end is not a valid fax device. ITM also supports email integration for user notification, fax server features, high volume fax broadcasting services, and transaction or entire-job management.

Faxing from the Web

Its also possible to request the same fax-on-demand documents via the Web. Ibex offers a Fax-From-Web option which allows Web surfers to have documents faxed to themselves or others. Its most commonly used in call center or sales environments where agents can request documents to be sent to callers. By using the Web, the sales or support agents can be located anywhere, and all have access to the same document library.

Email-on-Demand

Ibex offers an Email-on-Demand option which allows the same documents in the library to be requested by email. It works very much the same way that fax-on-demand does—a catalog of documents is available or sent on demand to the user. The user then selects a document which is immediately sent back, in this case via email.

An email address is published, for example info@abc.com, which users can email to. When the Ibex systems gets an email request, it sends back a catalog of documents. The user can then select a document and hit "reply" to send a response back. The Ibex system then retrieves the document and sends it via email or fax, depending on what the user selected.

Voice Recognition

Although not used in the USA, voice recognition (VR) is gaining popularity in Europe and South America because of the lack of touch-tone capabilities in many of these countries. FOD is an excellent application for VR because the voice application is quite simple and only number recognition is required. Speak a few numbers to select document(s), speak your fax number and you are done. There is usually no need to recognize letters or words, which is where VR starts to fail in other applications.

Service Bureau Applications

Service Bureaus own FOD equipment and charge others to use it. They typically push the system in areas that CPE (Customer Premise Equipment) customers do not. Ibex systems provide service bureaus with the ability to run large number of applications on the same system (usually based on DID number), account number access with database tracking, on-line or batch credit card charging, and high volume broadcasting.

For more information, visit: www.ibex.com

Chapter 33

Enterprise Fax-On-Demand

with FaxBack

For nearly a decade, FaxBack, Inc. has been the pioneer of high-performance fax-on-demand solutions. In 1997, with the acquisition of Instant Information, Inc. (I3), a leading fax and email development company, FaxBack has become the only computer-based fax vendor with its own 32-bit fax product line spanning FaxBack brand fax-on-demand systems, NET SatisFAXtion fax server software, CAS-based fax toolkits and fax driver software.

The Beginning

From its inception in the late eighties, FaxBack has played a leading role in the fax-on-demand market. The "faxback" name has long been synonymous with automated, interactive fax delivery systems. Chances are that when you dial a company for technical support or request literature, a FaxBack brand fax-on-demand solution stands behind your request.

Up until the acquisition of I3, FaxBack had specialized exclusively in developing and selling computer-based, turnkey fax-on-demand systems. The FaxBack brand fax-on-demand product line is marketed direct to corporate buyers as well as through international resellers. The company has had the most success selling direct to large service bureaus and corporations — which explains why currently over 30% of Fortune 500 companies are FaxBack fax-on-demand customers.

How FaxBack Fax-on-Demand Solutions Work

FaxBack fax-on-demand solutions are completely automated fax response systems that guide callers through a set of pre-recorded voice menus. Using a touch-tone telephone, callers select the documents or catalogs (indexes of stored documents) to be faxed back to any fax machine that is convenient to them...anytime...anywhere.

FaxBack fax-on-demand solutions can operate as standalone systems or they may be integrated to become a key part of today's business systems—bridging PBX, network, Internet site and/or host mainframe to the outside world. The documents callers request are retrieved from data files on FaxBack's local drive, a logical drive on a Novell or Windows NT network, a World Wide Web site, or from a third-party application using FaxBack's Application Programming Interface (API). These systems run on PCs and include a combination of voice cards, fax cards and software.

The Internet's Impact on Fax-on-Demand

Initially the explosion of the Internet had a negative impact on the fax-on-demand market in 1995. However, the sheer volume of information available on the Internet has contributed significantly to the demand for fax-on-demand products. And the company successfully capitalized on the convergence of its fax-on-demand technology with the Internet when it first introduced a series of fax-on-demand/Internet-integration products in 1995.

Fax-on-Demand Segmentation by Vertical Industry

FaxBack has taken a vertical market approach to help pinpoint the buyer of fax-on-demand systems. Significant vertical market segments for FaxBack brand fax-on-demand systems have included travel and tourism, government agencies, real estate, help desk, manufacturing, universities, purchasing and procurement, and Web page providers.

Implementation of FaxBack fax-on-demand systems varies by vertical market. For government agencies, like the IRS, it is an information dissemination fulfillment service. For the travel and tourism industry, it is a marketing and literature fulfillment tool. For realtors, it provides instant information on residential and commercial properties. For technology-savvy hi-tech firms it can be a technical support tool—helping callers answer their own questions without taking the time away from a relatively high paid technical support specialist.

Initial early adopters of fax-on-demand technology were primarily technology-savvy high-tech companies. More recently, fax-on-demand applications are being deployed in less technology-savvy environments including manufacturing, purchasing, and real estate.

FaxBack Fax-on-Demand Benefit Highlights

FaxBack systems offer a wide range of powerful benefits including:

- Unattended, 24-hour a day information services for customers and prospects
- Fast, return-on-investment, typically less than 9 months
- Reduced labor time associated with information fulfillment
- Accessible by anyone with a touch-tone telephone and fax machine
- Customers receive information at the peak of their interest
- Reduced printing, handling, mailing and overnight delivery costs
- Powerful marketing and sales tool for lead generation activities

- Frees technical staff by offering answers to frequently asked questions
- Seamless integration for fax-on-demand, LAN faxing, fax broadcasting and Web faxing in a single platform

A New Fax-on-Demand Landscape

The past 18 months have proven tumultuous for the relatively niche fax-on-demand market. As the fax-on-demand industry grew, so had the needs and level of sophistication of the fax-on-demand customer base.

Fax-on-demand vendors, increasingly found that in order to survive they had to offer tight integration with voice, fax, and document management technologies and provide seamless Web integration as part of their document strategies—all while continuing to invest substantial capital in research and development (in what was really still a relatively small market).

Beginning in mid-1996, a number of trends developed that found fax-on-demand vendors, following one of several paths:

- Acquisition by a fax server company,
- Formation of technology partnerships with interactive voice response (IVR) vendors and call centers, and finally
- Simply disappearing from the market altogether.

The FaxBack Strategy

The company recognized the fax-on-demand market trends, and in a strategic move to significantly expand it's product line, acquired Instant Information, Inc. (I3) in November of 1997.

I3, was the industry's premier fax and email development company. While a virtual unknown player in the everyday end-user vernacular, the past 15 years had found the company building an impressive resume with a variety of leading fax software and hardware companies.

The newly formed company promises to provide corporate users comprehensive, innovative and cost-effective solutions for a range of high-performance fax, email and voice applications.

The result today is a leading fax technology powerhouse that has become the only computer-based fax vendor with its own 32-bit fax product line spanning FaxBack brand fax-on-demand systems, NET SatisFAXtion fax server software, CAS-based fax toolkits and fax driver software.

For more information, visit: www.faxback.com

Chapter 34

Adding OMR/OCR to Fax Applications

with Cardiff Software

Reduce Data Entry Costs With Automated Forms Processing

People fill out forms to order merchandise or services, subscribe to magazines, attend seminars, request information. Virtually every company, large or small, relies on forms. Employees spend hours printing, filing, tracking, reproducing and ultimately getting rid of forms. It seems that every business transaction that applies to more than one person requires a completed form.

Handling traditional printed forms is an expensive proposition. Forms must be designed, printed, stored and finally manually processed. Forms processing is the lifeblood of business.

Data Entry Lags Behind Data Processing

Computer advancements have accelerated at dizzying speeds. Data processing tasks which 20 years ago required a building full of equipment and a multi-million dollar investment can now be accomplished by a single microchip on a PC. Similarly, data storage has undergone exponential cost benefit improvements.

While developments have served to improve data management processes, the function of collecting information-getting the data into the computer-has remained relatively unchanged. In fact, the faster processing speeds and larger data repositories have actually fueled the need for data entry clerks. According to the U.S. government, the data entry industry grew almost 50% between 1983 and 1991.

As a company grows and conducts business with a larger circle of customers, prospects, and vendors, more labor is required to feed the data processing engine. Data entry is a uniquely labor intensive piece of the business process. In fact, Spencer & Associates, a leading industry consulting company, estimates that service bureau staff and in-house workers spend over 31.4 million hours a year entering data.

The fact that data entry has remained primarily a manual task is an anomaly in the computer model. Each year, computer processing speed doubles, while the ability to input data remains relatively stable, thus creating the need for more labor in order to leverage the advancements in data processing technology.

This imbalance is beginning to change. Computer Automated Data Entry is revolutionizing data collection, reducing manual key entry—the hidden cost of doing business.

What is Computer Automated Data Entry?

Computer automated data entry is also known as forms processing. It is a method of accelerating data collection into a database in the most accurate and cost-effective manner possible. A computer automated data entry system automatically reads faxed or scanned forms and performs recognition on machine print, hand print, optical marks and bar codes.

The key to a successful automated system is a single system which accommodates all the major data capture vehicles: existing forms, controlled forms—forms optimized for recognition, electronic, and Intranet/Internet forms. With computer automated data entry forms are simply user-friendly vehicles for organizing data.

Computer automated data entry systems improve productivity and substantially reduces labor intensive manual data entry and forms processing by:

- Substantially reduce data entry personnel
- Eliminate paper documents and handling costs
- Improve access to mission critical information

More Than Eliminating Paper

Keep in mind that forms processing is really automated data collection. The form is only a vehicle for organizing the information. Effective computer automated data entry systems organize information in a way that makes it easy for recognition engines to read.

The key to computer automated data entry is accelerating the progress of information, no matter the format-paper, electronic, or HTML-from the user into a database. In the final analysis, eliminating paper is a side-benefit of computer automated data entry, not its main purpose.

How are Companies Using Computer Automated Data Entry?

Organizations are incorporating computer automated data entry into a wide variety of applications. Airlines and travel agencies are using automated forms to book reservations. Insurance companies are handling requests for quotes and processing claims with computer automated data entry.

Software companies are using computer automated data entry programs such as Cardiff Software's Teleform to collect product registrations and beta testing response.

1. Sales orders
2. Payroll (time cards)
3. Opinion polls
4. Standard test administration
5. Management reports
6. Clinical trials
7. Document distribution
8. Database entries
9. Remote computer operation
10. Insurance claims
11. Product registration
12. Advertising Response
13. Travel Reservations
14. Medical lab testing results
15. Index following images
16. Time and billing
17. Survey and poll taking
18. Tracking sales lead follow-up
19. Database access
20. Credit approvals
21. Fax tax filing
22. Government permits
23. Utility Bills

Fax Support for Computer Automated Data Entry

Facsimile is one of the most indispensable tools in day-to-day business operations. Very few businesses can operate without a fax machine to facilitate and enhance correspondence. It is estimated that by 1998 approximately 176 billion pages of plain paper faxes will be transmitted.

Most businesses use the fax machine and fax modem as an alternative to telephone and mail communications. Businesses realize that if a form is mailed in, there is postage to pay, envelopes to purchase, mail to sort, envelopes to open, not to mention the hourly wage or salary to pay. If the data is to be scanned in, expensive dedicated scanning equipment must be purchased. However, if timeliness is important, a faxed in form can input data into a database almost instantly.

With over 60 million fax machines installed worldwide, forms processing software that operates with fax images will expand fax usage from a correspondence medium to a cost effective platform for office automation.

In the past, there had been a strong distinction between forms processing software for scanner based applications and fax based applications. Most forms processing software was designed to read high quality 300 x 300 dpi (dots per inch) images from flat bed scanners. While the distinction appeared negligible, many forms processing applications failed to accurately operate with poor quality fax images. Today, sophisticated forms processing software, such as Cardiff Software's TELEform, is designed to overcome the differences between faxed and scanned images.

Fax machines normally input images at 100 x 200 dpi, as well as introduce skew, noise, elongation and black lines to the image during a transmission. TELEform will correct the incoming image, making the distinction between an incoming faxed image or scanned image nominal.

Before the forms processing software reads various entries on the incoming fax image such as hand printed, typed or shaded mark areas, the image is manipulated to correct for deficiencies. Let's look at the clean up TELEform performs for the fax image:

- **Deskewing** – Unlike a flatbed scanner, a faxed document is fed through a narrow reading element on a fax machine. The resulting image can be tilted or even upside down. To optimize the recognition process, the incoming image is placed in an orientation that is right side up and straight.

- **Line Removal** - Poorly maintained fax machines may have dirt in the reading element. This introduces dark vertical lines down a faxed image. Should the line run through a location where a hand printed or typed character resides, a misread will result. Removing this line, while preserving the original entry, is important to accurate form processing from fax images.

- **Despeckling** - Interference or noise caused by a poor phone connection during a fax transmission may introduce small speckles in a fax image. It is important that the speckles are removed, especially those that reside over an entry, for improved clarity.

- **Shrinkage and Distortion** - Based on the brand of fax machine, a fax image may be as much as (20 percent of the original length of the document. The fax image is reduced or expanded to its original dimensions to accurately locate entries on the form.

- **Field Detection** - Misfeeds and slippage may occur to a document fed through a fax machine. This will result in a non-uniform distortion to the subsequent form image. Since the software will be unable to determine the pre-defined location, field detection algorithms are used to overcome such difficulties. Sophisticated software will accomplish this task without timing marks on the page.

Once a fax image has gone through the steps above, the recognition results between a fax image and an image from a flatbed scanner are marginal. Computer automated data entry software that performs well on a fax image will operate well with higher quality scanned images, but the reverse is not true. If you plan to use fax and scanners to collect images for your forms processing application, be sure to pretest the system with poor quality fax images first.

It is important to note that Cardiff has worked closely with many LANFax and Fax-On-Demand software manufacturers and service bureaus. Therefore, TELEform easily integrates with a variety of the fax industry's most popular products.

As the next generation of fax machines arrive, the distinction between forms processing software for scanner based applications and fax based applications will become negligible. Group 4 fax machines transmit images at 400 dpi in gray scale and/or color, the same as today's scanners. Group 4 fax require digital phone lines called ISDN and are just becoming available in many large cities. ISDN lines are expected to be commonplace in the next 3 to 5 years.

As data communications continue to grow and the urgency for expedient information increases, the growing network of fax servers and fax modems is becoming invaluable. Combined with computer automated data entry software optimized for fax images, businesses can gain an affordable and immediate solution for processing sales orders, claim forms, surveys and more with the ubiquitous fax.

Voting Algorithms Improve System Performance

Any recognition engine, no matter how advanced, cannot be all things to all applications. Each has strong and weak points. Some are stronger at alphabetic rather than numeric characters; others work well on clean text but fail quickly on degraded text. Employing a single engine, while usually less expensive than multiple engines, results in greater user intervention for correction and validation. Taking a "one size fits all" approach to character recognition means lower recognition and accuracy levels, more flagged characters and lost productivity time spent in the verification mode.

In several major, independent studies voting algorithms produced superior results to single engine devices. The Information Science Research Institute (ISRI) at the University Nevada, Las Vegas compared the accuracy of six leading OCR products as well as a voting algorithm that made use of the very same engines.

With a test base of 132 pages containing over 275,000 characters ranging from clean to poor quality pages, the accuracy of individual engines was measured at 95.64% to 98.67%; the voting algorithm produced 99.33% accuracy. At first glance these numbers may appear comparable, until you consider the difference between the best single OCR engine and the voting process used by the UNLV researchers was over 15 additional errors per page - almost 2,000 characters.

In a recent comparison of handprint recognition (ICR) engines, renowned statistician Dr. Jon Geist concluded that a sophisticated multiple-engine recognition program produced substantially higher recognition rates than a leading single ICR engine. In fact, Dr. Geist's test showed that the voting algorithm resulted in 536% better recognition on upper case alphabetic characters, 350% better on combined upper and lower case alphabetic characters, and 241% better on numeric characters.

Multiple-engine recognition systems produce more accurate results than a single engine. The factual case for voting algorithms—the use of multiple OCR/ICR engines with a decision management layer—is even more compelling.

The fact is many other features, apart from engine performance, affect the accuracy of a. Pre- and post-processing, image clean-up and contextual logic are essential for forms processing accuracy. Yet these activities cannot completely compensate for weaknesses inherent in single engine schemes. Voting algorithms have the ability to apply intelligence and fuzzy logic to the recognition process, tracking engine performance for each task and using this information in future decisions, thereby leveraging the strengths of each engine.

Superior Accuracy Through Quality Form Design

Accuracy is the key to effective forms processing. Too often, however, the responsibility for system accuracy is shouldered by the recognition engine. Developers and customers alike demand increasingly sophisticated OCR and ICR functionality to reduce recognition errors.

The recent progress of recognition technologies, including multiple engines, neural networks, voting algorithms and fuzzy logic have led to substantially increased forms processing productivity. Despite these gains, it should be noted that recognition engines do not work in a vacuum. Recognition is merely one step (albeit an important step) in forms processing. Achieving maximum accuracy requires thoughtful design of the entire system, from forms design through post-processing operations.

Taking this macro-perspective further, we need to keep in mind that the goal of forms processing is to capture information. To this end, the most important issue affecting the success of a forms processing solution is not whether the OCR engine is optimized for forms processing, but rather that the form itself is optimized for collecting information.

Starting with forms that are designed specifically for logical data capture, whether hand print, machine print or marks, simplifies every other step in the process. Think of the engine as a detective looking for clues to identify each character or mark on a page. Regardless of the detective's experience & intuition, he stands a far greater chance of identifying clues if he knows (1) what they are, and (2) where to find them.A well-designed form can provide this information to the recognition engine.

Some vendors have approached the accuracy issue by attempting to cluster pre- and post-processing activities within the engine itself. Specialized forms processing engines bundle these numerous sub-routines. These efforts have met with limited success because PCs are not capable of the parallel processing required for sub-routines. For the most part, this approach has served only to complicate the recognition process and raise the cost of data collection.The combination of activities comes at a hefty price: typically a ten-fold increase in initial software investment, as well as similar increases in training and ongoing maintenance.

Another solution has been to use existing high-performance recognition engines solely for recognition. Solid coding is then applied to tune image clean-up, deskew, and contextual forms checking apart from the recognition step. Regardless of where the secondary activities take place, the basic recognition function remains the same.

It is far more important to integrate intelligent forms design within the data collection solution rather than integrating OCR/ICR engines with post-processing. Relying on the engine to cover up for poor overall system design neglects the importance of deskewing, indexing, and verification, while rendering incremental gains in recognition accuracy meaningless.

Summary

Computer automated data entry is a hybrid solution that spans the gap between personal productivity tools and departmental/organizational systems. The development of cost-effective software that leverages standard hardware platforms has led to data capture systems that are viable for small and mid-size companies, as well as Fortune 500 companies.

While computer automated data entry encompasses several familiar technologies — OCR, document imaging, facsimile, scanning — the goal of computer automated data entry is simple: reduce or eliminate manual data entry.

For more information, visit: www.cardiffsw.com

Section VIII

Hiring Enhanced Fax Service Bureaus

Section VIII

Introduction

Fax Broadcasting and Fax-On-Demand are powerful sales and marketing tools. How else can you reach thousands of your existing customers in minutes, for about the cost of a first class stamp or provide inexpensive and automated document fulfillment, 24 hours per day those without web access? Okay, you agree the applications makes sense, but what's the next step?

Usually, it's deciding whether to buy versus build the system. Do you contact system manufacturers and VARs, or call hardware, software and component makers to piece together what you need? There is a third, quite compelling approach: outsource.

Since the first one was launched in 1989, the Enhanced Fax Service Bureau industry has grown steadily in the US. Top fax bureaus have made large capital investments and employ the programmers and engineers necessary to impress, well, even me.

No longer do six ports and two guys working from their garage, equal a bureau. Today's bureaus are sophisticated operations, running dozens of T1s. Their fax broadcasting offers features like imbedded, multiple field personalizations, watermarks and intelligent retry functions. Fax-On-Demand has grown-up also, with features like on-the-fly document creation, ANI, live feeds into news and wire services and credit card capture, available. How long would it take you to build this stuff in-house?

And why would you, when enhanced fax service bureaus offer great alternatives? Consider this, when hiring a bureau there's no capital outlay and your start-up time is minimal. Going the bureau route also means there's no hardware to install and maintain or new software to learn and upgrade. Bureaus guarantee little or no down time and put no new pressure on your network. Since fax is their core competency, bureaus already have the staff to make your application a success. Why hire fax programmers, engineers and administrative help when the bureau provides them for free?

About 75% of all FOD applications and nearly half a billion dollars in fax broadcasting are managed and transmitted by bureaus. Think only small companies hire bureaus? He's a short list of those who outsource fax: AP, ARCO, CNN, DHL, Disney, Hilton, HP, IBM, ITT, Motorola and S&P.

There is a fundamental shift occurring in corporate America today. Organizations are beginning to realize they don't understand and can't control their fax costs. That is, enhanced fax and plain vanilla point-to-point faxing. So, they are hiring specialists who can. The fax industry recognizes this and is taking action. Competing fax server companies are partnering with competing fax service bureaus to offer their clients complete solutions.

Witness the recent announcements of strategic partnerships between leading fax server companies and fax service bureaus. RightFax and FaxSav announced a partnership and so have their rivals Omtool and Xpedite.

What's going on is, large, medium and even small organizations are installing LANFax servers which can manage and automatically direct fax traffic to service bureaus. The bureaus have the telco and/or IP infrastructure to manage and route fax traffic more cost-effectively. Besides also offering the enhanced services.

Kinko's convinced us to outsource our copying and the same is happening with fax. I predict, by the year 2000, over a million desktops will be outsourcing their fax traffic, and service bureaus and will be saving corporate America over a $100 million dollars per year.

Ready to put your toes in the ocean? The best reason to hire a bureau is your ability to try it, before you buy it. These questions should help get you started:

9 Questions To Ask Before Hiring a Fax Service Bureau

1) What enhanced fax services do you offer and do you specialize in any one service? Some bureaus (the RBOCs and carriers that offer fax) may specialize in fax broadcasting and actually outsource their Fax-On-Demand services to others. Make sure your bureau is competent in the fax technology you need. That it's not just a 900 bureau that thinks it understands fax.
2) How long has your bureau been in business? The first bureaus were founded about 1990, so the top bureaus have been sending faxes for five to seven years.
3) For Fax Broadcasting: How many one page documents can you broadcast in an hour? The only good answer, is the one that meets your needs.
4) For Fax-On-Demand: Will I have menu and script approval, before the application is engineered and prompts recorded?
5) What is your turn around time? For Broadcasting, one

day to a few hours is normal. For FOD, thirty days to live application is acceptable.

6) How do you charge: per minute or per page? (Hint: Per minutes, with six second billing is usually less expensive, however per page will be easier for you to budget.)
7) What are your set-up fees, minimums, storage charges and are there any hidden fees? There's as many ways to price fax services as there are long distance calling plans. Ask plenty of questions and don't be afraid to request a written quote.
8) What are the phone numbers of three Fax-On-Demand applications you currently maintain? If three won't be provided, call another bureau. However, once you have them, call them to sample the applications. Do you like them? Do they sound professional? How do the documents look and how long did it take for them to be transmitted?
9) Will you provide three client references? Again, without three, look elsewhere.

Advice from the author: Don't be cheap—you will get what you pay for. Like your company, fax service bureaus deserve to make a profit. They know what their costs are and what their competition charges. Try to play one bureau against another to shave a penny per minute and the best bureaus will turn down your business. The top bureaus are busy and smart, and refuse to play this game. Look to build a mutually beneficial relationship based on trust, not cost.

Chapter 35

The Enhanced Fax Service Bureau: Epigraphx

Introduction

Outsourcing: Getting the Most From a Fax Service Bureau

There's a lot more to the fax services market than there was a few short years ago. It's not just about Fax-On-Demand anymore, thanks to gains in fax technology and a general proliferation of interest in fax applications. What used to be known strictly as Fax-On-Demand is now often referred to as "enhanced fax services"—signifying a potpourri of applications for companies to choose from. Along with Fax-On-Demand, there's Internet-based faxing, faxing from the desktop computer, optical recognition fax forms that capture customer data automatically, dynamic merging of database information into individual fax transactions, and more.

While there's some excellent off-the-shelf fax software on the market today, it doesn't come with the expertise needed to tie fax services together into an enterprise-wide solution. Nor does it supply the staff time and inclination required to maintain and grow one or more fax applications. Enter an outsourced fax services bureau.

In fact, even many companies that do have the staff resources and technical prowess—F1000 companies like 3M, Apple, Hewlett-Packard, and Cisco—still choose to outsource automated fax services. Why? Corporate cost-cutting and a renewed focus on core competencies are driving the growing trend toward outsourcing. The primary benefits are appealing:

- **Substantial cost savings and no up-front risk**

  Many companies have implemented internal fax communications systems, only to turn to outsourcing when the applications became too complex, multi-departmental, and costly in terms of equipment and personnel. With outsourcing, there are no capital equipment or software costs, and literature fulfillment can be achieved at 50 to 70 percent the cost of traditional paper-based distribution. A fax services bureau manages every aspect of system capability and capacity, application design, service implementation, training, security, and data reporting. Or, a company may choose to supplement an in-house fax application with outsourced services. Outsourcing is also a cost-effective way to "test drive" fax-back options in sales and marketing campaigns.

- **Enterprise-wide implementation support**

  In corporate America, no one is clearly "in charge" of fax communication. Should Marketing Communications manage it? Telecom? MIS? Customer Support? Early fax communication adopters (companies such as Apple, Hewlett-Packard, and Intel) discovered that their fax applications were driven by departmental mavericks without consistency and control from computer or telecommunications managers. As fax applications proliferate, companies still struggle to determine which department should own and manage enterprise-wide fax applications. So they turn to an outside bureau to provide and manage enterprise-wide implementation support.

- **Automated customer data collection**

A subtler benefit of many automated fax services is the customer and prospect information they yield—including name, company, fax number, and history of information requests that divulge specific interests. An outsourced fax services bureau can ensure that a company's data is collected, sorted, and reported quickly—in time to follow-up on sales leads while they're still hot. Extensive fax databases, gathered over time, are valuable in developing future fax and print marketing campaigns, especially for products and services with sales cycles of several months or even years.

For these reasons, among others, the fax services business is flourishing, and the list of companies offering automated fax services is long and varied. Finding the right one requires a well thought-out assessment of your fax objectives, and careful evaluation of the capabilities and expertise of prospective bureaus.

First, the Groundwork

Take the time to define your fax application objectives and audiences before you start shopping for a fax services bureau. Defining your objectives and audiences up front will not only lay the foundation for a smooth fax implementation but will go a long way in helping you select a services bureau best suited to your needs and requirements. For example, your objectives may be to:

- Reduce fulfillment costs
- Shorten sales cycles
- Lower data-entry costs and shorten data turnaround

- Streamline lead generation
- Automate customer and technical support

Use your objectives to determine who you want to reach with fax and which areas of the company your application will service. Some fax applications are designed to field requests from varied audiences, while others are highly targeted or access-restricted. Prioritizing your target audiences will prove especially helpful during actual implementation, when you'll have to make decisions about user points of access, voice scripts, and other features of your fax application.

Think also about the scope of what you want to achieve—not just with Fax-On-Demand but other automated fax options as well. Do you want to link all your customers and staff through one centralized system, or are you simply augmenting your call center or other company department with a separate document request line? Do you plan to experiment with a variety of fax services as you test marketing campaigns? Are you looking for a bureau to manage an occasional fax broadcast or a long-term fax services partner to assist ongoing sales or support efforts? Are you thinking about fax-enabling your web site?

Fax Document Library: Hunting, Gathering, and Access

At the core of many fax applications is a centralized document library, a collection of the documents (in electronic form) that you wish to make accessible by fax. While the day-to-day maintenance and updating of your library is done by the fax services bureau, initial document hunting and gathering is an important step, and one that an outside bureau can't help you with to any great extent. This process inevitably will take you longer than planned. In deciding which documents to include, remember the 80/20 Rule—80% of your callers will request 20% of your documents. So, spend your energy locating the latest versions of that coveted 20%.

The procedure for submitting fax library documents varies, naturally, among fax services bureaus. Inquire about it and ensure that you're comfortable with the process. Document submission shouldn't be arduous or time-consuming, nor should you have to contend with a lot of re-formatting. A bureau should be equipped to accept your documents in virtually any format: Acrobat PDF, Quark Xpress, Microsoft Word, PageMaker, EPS, PostScript, G3 fax, TIFF, ASCII, PCX, and DCX. Hard-copy or electronic documents are usually acceptable; however, the highest quality fax output is derived from electronic files.

Whatever your library includes, look to the fax services bureau to help you group documents so that callers can easily locate and request the information. Expect an outside bureau to take about two weeks to organize and implement a 50-document library.

User access to your fax document library can be achieved in several ways, depending on your current information distribution methods and the extent you wish to supplement them with a fax-back option. Points of access include:

- Existing corporate 800 number
- Separate, dedicated 800 number
- Toll number for all callers, including international
- World Wide Web (via an Internet fax gateway)
- Fax gateway from the desktop computer
- OCR/OMR forms

Find a services bureau that can integrate multiple points of access into your document library. You may not be ready to fax-enable your web site today, but if it's something you envision down the road, a services bureau should be able to add World Wide Web access to your document library smoothly and efficiently. In other words, make sure that the firm you choose has the expertise and flexibility to grow with you.

Fax-On-Demand (800 or toll number access). If you want your application to work off an existing number that your customers are familiar with, integrating Fax-On-Demand with the number makes sense. If customers are accustomed to your specific call structure, there's no need to separate your Fax-On-Demand option. Integrating does take more of the caller's time to navigate to a specific destination. Most Fax-On-Demand applications added to an existing corporate 800 number menu are for document retrieval from departments such as technical and customer support, marketing communications, and investor relations.

But if your callers must navigate multiple options (more than five) or multiple menus to find the information they want, a separate Fax-On-Demand application may be your best approach. Separate 800 or toll numbers are typically the way to go for companies that have several product lines with multiple products, widely varying product lines, or products sold to diverse industries. For instance, if your company sells microwave tubes in the communications industry and power supplies in the medical equipment industry, individual Fax-On-Demand applications would be most efficient—each delivering the detailed, highly targeted information required to reach these very different audiences.

Also, if you are advertising a new product and want to fulfill routine information requests via fax, a stand-alone service will prevent caller confusion. Callers won't have to navigate multiple levels of information from a single corporate 800 number used for other purposes or products unrelated to your new product. In any event, a fax services bureau should be able to help you sort out which Fax-On-Demand integration—including call flows and document indexing—is right for you.

Other areas to explore are voice response and custom Fax-On-Demand capabilities. What languages are supported, and are they available in both female and male voices? Is voice scripting provided? What's the turnaround time for changing voice prompts? Does the fax services bureau offer customized features such as Automatic Number Identification (ANI), Personal Identification Number (PIN)-coding, outdialing, and online credit card processing?

Fax-On-Command (fax gateway from the desktop). Employees have direct access to your fax library from their desktop computer. This application is commonly used in telesales and technical support/customer service environments—minimizing the amount of time agents spend fulfilling requests for routine information such as spec sheets, tech tips, product updates, and dealer locations.

Internet Fax-On-Command (World Wide Web access). Employees and customers with Internet access can go to your company's web page, select desired documents, and send them to any fax machine. Also good in call center environments, Internet Fax-On-Command has the added sales benefit of automatically capturing the name, company, and phone number (at a minimum) for each person requesting web information via fax.

OCR/OMR fax forms. Using Optical Character Recognition (OCR) and Optical Mark Recognition (OMR), these forms capture data automatically via fax. OCR/OCM forms are used instead of a telephone keypad to retrieve documents from an electronic library, or they can be used solely to capture customer data. Included as a response mechanism in direct-mail campaigns (vs. a Business Reply Card) and print advertising, OCR/OMR forms invite recipients to answer profiling questions and choose from a selection of faxable documents. Scannable data points are programmed to fulfill their literature requests without human intervention. Responder profiling information is automatically captured to a database.

A services bureau may also provide fax applications not linked to a centralized document library, including:

- Fax broadcasting—The near-simultaneous transmission of a document to many locations. Used for distributing product announcements, press releases, newsletters, registration forms, event confirmation, etc. Some service bureaus also handle email broadcasting.
- Dynamic fax merge—Generates, merges, and collates

custom faxed documents based on data input (the same type of documents typically generated by traditional mail-merge). Used for purchase orders, invoices, account balances/portfolio updates, order status, event confirmations, registration packets, etc.

- Fax routing—Enables users to view, save, print, forward, or delete their faxes from anywhere they have access to a web browser.
- High-volume fax receive—Provides a never-busy, toll-free response mechanism for high-volume incoming faxes. Used for customer-record deletion requests, address changes, etc.

A Look Behind the Scenes

In addition to finding out about specific fax applications and their implementations, you'll want to examine a fax services bureau on several "behind-the-scenes" fronts. Chief among them are system modularity and transmission speed, system monitoring and backup, account management, and reporting capabilities.

System Modularity and Transmission Speed

The number of incoming line and outgoing port capacities a bureau has is, essentially, a non-issue for companies who choose to outsource. Any fax services bureau today with state-of-the-art equipment can expand its capacity in less than a month. A more telling question to ask a prospective bureau is, "How does your infrastructure support your clients' needs, and how do you determine when increased capacity is needed?" Ensure that the services bureau has a protocol for monitoring volume capacity.

Fax transmission speed is important—especially since bureaus charge for Fax-On-Demand and most other fax services on a per-minute transaction basis. Seek out a bureau that can transmit faxes at the highest fax speed, 14.4 kbps. In this day and age, there's no reason for a fax services bureau not to at least attempt to send its transmissions at 14.4. At that speed, a page of text takes 45 seconds to 1 minute to transmit. Don't be shy to ask the bureau for recent statistics on its transmissions times vs. speed. What's more, a bureau should be equipped to adapt to faster transmission speeds as technology advances and fax device speeds increase.

Also, find out whether you'll be charged for actual transmission time (i.e., only for documents delivered successfully), and not for delivery attempts that fail because the receiving fax is out of paper or busy. What's the bureau's process regarding fax retries (how many tries are attempted before the transaction is classified a failure)?

System Monitoring and Backup

Around-the-clock monitoring of telephony networks, computer networks, and outbound sending systems is critical. Your fax services bureau should have an automated process for testing every aspect of its system at tight intervals (e.g., every 15 minutes), and assurance that a technician will respond immediately if a malfunction is detected. Servers and hardware should be rack-mounted in a climate-controlled, window-less room (the minimum protection against earthquakes, windstorms, and other uncontrollable rumblings).

In addition, ask whether the system is protected by an uninterrupted power supply (UPS). What about a back-up generator to further guard against power failures and surges? Are there redundant servers? Does the fax services bureau have a formal procedure for data recovery in the event of a power outage or disaster? If so, when was it last tested?

Expect a bureau to perform nightly backup of all its systems and databases. There should also be routine (preferably weekly) backup of all system files to tape or some other means, for storage in a secure, off-site facility.

Account Management—a Team Effort

In the area of account management, it's important to know what you want and expect from a fax services bureau. Account management can vary dramatically for a small company vs. a large one. Does the fax services bureau have experience dealing with companies the same size as yours? Who will be on your "account management team" and how will they interact with you? What procedures (formal and informal) and systems are in place to meet your needs and educate your company on various fax applications and uses?

Regardless of your company's size, you should feel confident that there will be enough staff on the bureau's side to respond to your questions and concerns on a day-to-day basis (if need be), as well as expedite new fax requests, changes, and updates quickly.

Remember that the fax services industry is built on speed and efficiency—your outsourced bureau should be attuned to the notion.

Production. Depending on the bureau's breadth of expertise, production services may include voice script modifications, document creation, conversion and image enhancement, cover sheet creation and/or modifications, database management, and application modification.

Since most documents aren't designed for maximum fax transmission or visual fax impact, you'll likely want to query the services bureau about its conversion and resolution enhancement techniques, including acceptable file formats. Be assured that your documents—text and graphics—will be optimized for high resolution, readability, and speed in fax transmission. For instance, to attain high quality and depth for the fax medium, documents can be converted to 200 dpi. Background screens can be removed, and color stripped and alternatively presented in black, white, and gray scales. Commonly converted documents include product brochures, catalogs, technical notes, data sheets, maps, press releases, and ad slicks.

Another point to consider here is your own corporate document standards. How important is preserving the integrity of your company logo, trademarks, or other corporate emblems in documents transmitted via fax? If this is a weighty consideration, look for a fax services bureau that's respectful of your corporate standards and procedures—and can heed them without compromising document turnaround.

Technical operations. Again, depending on the bureau's breadth of expertise, operations can encompass technical support, custom web page development, and custom reporting and programming services.

Marketing. A handful of fax bureaus have demonstrated knowledge of integrating fax services into a company's mass communications strategies. It's worth your time to seek them out if you want guidance in how and when to deploy fax-based marketing efforts such as OCR/OMR forms and lead-generation fax surveys.

Reporting Capabilities

Tracking fax application usage is an essential task, and a services bureau should provide you with regular usage summary reports (weekly for Fax-on-Demand and Fax-On-Command, and at least monthly for other services). Usage summaries track the number of customers who access your fax service, how much time they spend using it, which fax documents are requested, and fax delivery success and error rates. This type of reporting not only helps you keep tabs on your fax application activity, but is a handy resource for database list clean-up and lead information.

What type of data will you get back? Reports should be easy to read and usable—you shouldn't have to work hard to find application usage data such as:

- Total number of calls received during the period
- Inbound and outbound call duration
- Documents requested—in order of fax request priority

- Fax delivery error summaries—how many faxes were attempted, failed, and why

Ask how the reports will be delivered to you: hard copy, electronically, both? Reports in electronic form (vs. hard copy only) will enable you to perform further data manipulation, if desired, without duplicating data entry. Also, find out whether the bureau can generate custom reports (e.g., by ANI or PIN data, or by geographical region).

References, References, References

Whatever you do, don't choose a fax services bureau based solely on pennies per page, transmission speed, or any other single cost or performance criterion. Ask for client references, three at least. Specifically, get the names of people for whom the bureau has developed and implemented the same application you are seeking.Also ask for a fax demonstration-the phone numbers of actual applications you can access, listen to, see, and sample yourself.

- Are the voice prompts intuitive and easy to follow?
- Does the voice sound professional—one you want to represent your company?
- Is the document you requested legible? Does it look good?
- Is web access to fax documents clear and easy to maneuver?
- How long did it take for the requested document to transmit?

And, lastly, have some fun. In choosing to outsource fax services, you've already made the decision to expand your customer reach and improve mass communications with proven technology—at low cost and no risk to you. Finding the right fax services bureau should be accomplished with forethought and care, but it's simply the next step in meeting your larger, important communications goal.

For more information about hiring enhanced fax service bureaus or to learn more about Epigraphx, visit them at www.epigraphx.com

Chapter 36

The Internet Fax Service Bureau: FaxSav

History

FaxSav was founded in 1989 dedicated to providing state-of-the-art productivity enhancing, cost-effective faxing solutions to businesses. Its original network was telephony based and customers accessed it from their fax machines using a dialer device called the FaxSav connector. The network provided productivity benefits through smart retry schedules and cost savings for international faxes through an optimized high-quality, digital network and pay-only-for-successful faxes pricing model. The service, called FaxSav Plus, continues to be popular today, especially with companies with a high degree of international faxing where it is still difficult to get faxes successfully delivered through traditional telephony networks. Other attractions are its options for confirmed delivery and activity reports. For instance, delivery reports which contain a compressed image of the first page of the sent fax are returned to the sender, and are often kept on file by some customers as proof of delivery.

Broadcast Services

FaxSav's next major product introduction was the industry's easiest-to-use, flexible, and powerful broadcast service. Called EZ-List, the service has appealed to heavy broadcast users such as computer hardware and software distributors, trade show promoters, PR companies, and any vendor wishing to use fax as their document distribution and communication medium. Users can initiate broadcasts with their source document either in FaxSav's Internet-enabled fax software FaxLauncher (discussed later) or from a standard fax machine with a FaxSav Connector. In 1997, FaxSav introduced List Manager software for customers wanting to control and update large broadcast lists on-the-fly. The software has been found to be easier to use, yet more flexible in importing capabilities, than other similar tools.

EZ-List is just as attractive to "low-volume" users because of its ease of use. From a fax machine a document can be broadcast to hundreds or thousands of destinations with the press of just 3 buttons on the fax machine. PC users also find the FaxLauncher software extremely easy to use from any Internet-connected desktop (Macintosh or Windows). Moreover, initiating a broadcast from the PC allows the highest image quality possible.

Integrating the Internet into the Network

FaxSav began its evolution into an Internet fax company in 1996 by integrating the Internet as a transmission medium. The company began deploying Internet fax nodes in major countries around the world to avoid intercontinental telephony costs and to increase quality and performance. Indeed FaxSav was the first network company to deploy Internet fax nodes, and continues to be the leader in terms of coverage. By 1998 nodes were operating in 21 countries covering over 60% of the world's telecommunication destinations based on volumes. While the company continues to add other countries, the network's telephony delivery systems remain in place to reach every location in the world. Most importantly the network fully integrates the two technologies so that in the event that an Internet node is unreachable due to Internet congestion, for instance, faxes are transparently re-routed over telephony with no disruption or price premium to the customer. Any time faxes are on the Intranet portion of the FaxSav network, they are RSA-encrypted.

Reliability and accountability are the hallmarks of the FaxSav Global Internet Fax Network and the company's suite of faxing services. The network is engineered for 100% availability with complete redundancy of all its components. Nodes are all located in major telecommunications facilities that provide extensive power backup capabilities. All of the company's Internet fax services are termed virtual real-time meaning that the time difference between a fax coming into the network and the destination fax machine first ringing is less than 15 seconds. As the first fax network company to successfully integrate the Internet into its technologies, FaxSav stays at the lead in providing dependable services for companies relying on it for their critical fax communications.

Desktop-based Internet Products

FaxSav began launching its desktop-based Internet faxing services in early 1996 with the first being an email to fax service called FaxMailer. This service lets a customer use any email software, from any platform send email messages to fax machines. As of 1997 the service supported primarily text messages and several graphic file formats. More popular word-processing and spreadsheet formats, for instance, will be supported in 1998. As for all of its desktop-based services, delivery notices are returned to the sender's email address.

Only a couple of months later FaxSav introduce the first Internet-based desktop faxing software, FaxLauncher, which enables PC and Mac users to fax any document in their computer — including graphics and other complex documents. Additionally the software RSA-encrypts faxes as they are sent through the Internet to the FaxSav network. FaxLauncher is available in Light and Pro versions with the Pro offering additional features such as a facility to resend failed faxes and track fax progress in real-time through the FaxSav network.

Although the services mentioned so far are the most popular, the company has an even wider range of capabilities, unmatched by any other fax vendor, Internet or non-Internet. FaxScan software is available for paper-based faxing in countries where FaxSav Plus service is not yet available. The software enables owners of TWAIN-compliant scanners and Internet-connected PCs to operate their scanner as a virtual fax machine.

FaxProxy is a service for fax-enabling Web sites. No additional hardware or software purchase is required. Simple CGI-scripting and special provisioning in the FaxSav Network is all that is necessary to implement FaxProxy. For instance any company doing any type of Web commerce can utilize the capability to transparently route communications, orders, and transactions from Web visitors to fax machines anywhere in the world. Another popular application is fax-on-demand from the Web site of company and product information, often implemented as a customer-service operated Intranet application.

Another unique offering from FaxSav is its ability to work with a customer on a custom fax application. In any instance where a customer has a system application that generates data that needs to be distributed via fax, FaxSav custom applications development staff can work with a customer's development staff to deliver a solution. By virtue of FaxSav's flexibility in receiving information from the customer whether it is data to be merged with a graphic template, or the customer's application is generating the fully formatted fax image, there is a possible solution which requires no additional hardware on the customer's part and takes advantage of the Internet to provide ubiquitous ingress.

The latest of the additions to FaxSav's faxing suite is FaxCourier, a fax to email service that completes the loop in a total virtual fax server solution. Unique local and toll-free fax numbers are available. Senders fax as usual to these numbers, and the network offers never-busy receipt. Faxes are converted to TIFF file format and emailed to the customer's email inbox as a MIME attachment.

Technology Partners

Recognized as the leader in Internet fax networks, FaxSav's technology and universality of access has attracted several technology partners who team with FaxSav to offer unique solutions. SmithMicro's end-user oriented HotFax software was the first to include FaxSav's ServerLink Internet faxing components. Hewlett-Packard integrates its Network Scanner with FaxSav that turns a share-able scanner into a high-capacity virtual fax machine. In addition Hewlett-Packard bundles FaxLauncher software with its line of flatbed scanners.

RightFAX, a leader in the fax server market, integrates ServerLink so that their customers can leverage FaxSav's limitless network capacity and global coverage no matter how many servers are installed on the customer's premises. Ericcson, the cellular communications company, bundles FaxMailer with its palmtop computing and communication devices.

Market Adoption

As of 1998, over 40,000 corporate customers were depending on FaxSav for their fax communication needs utilizing one or more of FaxSav's services. More and more companies are turning to FaxSav to outsource their faxing requirements as they consider whether they themselves want to be in the fax business, or instead, concentrate on their core competencies.

Customers range from Fortune 500 multinationals implementing FaxSav as the enterprise-wide desktop faxing solution to small businesses located in remote parts of the world with just a few PCs. Indeed FaxSav's ability to provide services to locations all over the world set it apart from many other new entrants to the market.

FaxSav has consistently been recognized as a leader in technology and quality year after year, and was awarded in 1997 the Best Enhanced Fax Network. Other external testimonial to FaxSav's excellence is in the degree of interest in reselling relationships. FaxSav now has over 150 selling partners in over 50 countries around the world.

For more information, visit: www.faxsav.com

Section IX

Important Faxing Considerations

Chapter 41

Faxing Internationally

Introduction

Facsimile technology and its widespread use is not an American phenomenon. It is a valuable business tool, in use worldwide. As such, the system integrator and user doing business internationally, must be certain that those fax devices and fax systems, that are vital links to customers (and their customers) across the globe, not only work as promised, but are fully compatible electronically and legally with the public telephone networks of the countries where these systems will operate. The only way to ensure one hundred percent compatibility with the public telephone networks of other countries, is by having the main components of these systems, in this case CBF boards, officially approved, or homologated by the government authorities in charge of public communications in each country, (generally the Post Telephone & Telegraph (PTTs) administrations.)

When a system integrator (and customers using the system) uses a CBF board that has not been homologated, by the country where it will operate, the risk is run that subtle electronic incompatibilities between the unapproved CBF board and the foreign public telephone network may crash the system and even damage the public network. Less serious results can be higher telephone bills due to incompatible signal transmit levels, receiver sensitivities, and other factors that may cause the system to do early disconnects (wasting calls), or prolonging calls by forcing it to fall back to considerably slower transmission speeds.

Not every CBF manufacturer, however, pursues homologations. The process is expensive, time-consuming, and requires inordinate patience. Some manufacturers, have recognized that because fax is a universal technology customers must have solutions that cut across international borders.This requires that the manufacturer create and maintain a credible presence in the world market. Before this can become a reality, however, it becomes necessary to comply with the different requirements and specifications for communication products, deemed essential by each nation.

Without these international approvals no manufacturer can assure customers that the systems they create and use will work in the countries where they operate or, what is more basic, that those nations' authorities will even allow these systems to be hooked to the public telephone network.

Standards are Not Universal

Although the International Electrotechnical Commission (IEC) has established comprehensive guidelines for areas such as operator safety, there is no one widely accepted standard for safety protection for modems, CBF boards, and other telephone apparatus. Model rules may be drawn by international bodies and be agreed to in principle by the various countries, but they have not been implemented wholesale at the various national levels.

Thus, safety tests for CBF boards differ from country to country. The tests are carried out differently, the voltage requirements are different, the way the equipment is tested for overloads is different, the number of seconds the device under test is supposed to withstand the breakdown is different, and so it goes.

While coping with differences in telephone networks (such as one exchange needing pulse dialing with a 60/40 make/break ratio, while another requires 67/33) is not a major problem, other requirements can become quite complicated. Again, safety regulations are a good example. Although it would appear lightning strikes telephone poles in the same way in Finland, France and Germany, and cars knock down power lines on to telephone wires in the same way in Italy as they do in England, rules differ.

As with the United States, experiences with these situations have been unique for each country and the laws and regulations that have been established as a result reflect this. Testing procedures differ as well. Although European Economic Community (EEC) rules now theoretically apply equally across Common Market nations, standardization is yet to take place. It will be some time before it is possible to get a certificate out of a testing lab anywhere in Europe and have it cover equipment throughout the EEC. The general adoption of uniform EEC standards, especially for analog equipment, still appears far in the future. A homologation granted in the UK will not be good in Spain any time soon.

It is not surprising then that just collecting specifications needed for compliance can be a major, very expensive, undertaking. A veritable profession has developed around this need, and consulting firms specializing in the gathering of this information for companies seeking homologations have appeared all over the world, offering expertise of varying value and depth.

The Homologation Process

On the average, collecting information on the various specifications that must be met to receive homologation can range anywhere from 30 days to six months. This process not only involves gathering the information, but also having a native sponsor, a local address, contracts authorizing the people who are going to sell the product to act as your agents in the country, and other issues that vary from one national jurisdiction to another. None of this can be worked out from the United States. A local representation is a must. Even if all specifications are obtained and a prototype that follows them scrupulously, is then produced, by the time the product begins going through the approval process itself, it is not unusual to discover that in the interim there has been one or more bulletins invalidating or adding some of the parameters used to build it.

Technology does not stand still, and specifications and safety regulations evolve continuously to keep pace with these developments. This not only occurs in hardware and technical situations, but also in the more complex areas of protocols. The level of finished product that a country requires before certification is granted is also a variable. In some countries an absolute production-level product must be presented and no substitutes are accepted.

Other countries are satisfied with a one-of-a-kind prototype and will grant approvals based on that. Sometimes protocol testing is part of the approval process sometimes it is not. The situation has been encountered where the approval agency is solely interested in the telephone interface power levels and safety of the CBF board, and not in its actual operation as a fax device, or in its capability to communicate with other machines. Often, part of considering a product as being complete for the approval process is meeting the requirement for a full translation of all material and manuals connected with the product, including software screens. Generally, however, it is only required that user documentation be translated.

On the average, it takes six months to produce a prototype ready for preliminary testing at the target country. At this time, a contract is entered into with an approved testing laboratory in the target country that (just like its FCC-licensed United States counterparts), is authorized by the pertinent ministries to issue compliance reports. The CBF board prototypes are shipped, and the approval process then begins in earnest. The first result is a very detailed preliminary report outlining where the product must be improved or why it did not pass.

Once the preliminary report has been received and studied, it typically takes six to eight weeks of additional work to implement changes, tune up circuits, re-parameterize the software, deal with all the issues that were missed, and come back with another product iteration. A second contract is signed with the testing laboratory, and two to three more weeks are required to put the product through its paces again. Everything is retested: software, hardware, and circuit design. If the necessary changes will be reflected in the manuals, new translations of these must also be provided.

During this testing process it is worthwhile for the company seeking homologation to have at least one of its engineers present at the official laboratory, to ensure that the installation and use of the product are within specifications. Since the laboratory technicians do not have the time to learn the product, it is advantageous to have in place someone knowledgeable to assist with the installation, set things up, and then withdraw into the background ready to reappear if assistance is needed or further questions need answering. While the engineer's presence does not affect how the laboratory measures the product, it helps the process along by ensuring it is done correctly.

Finally, if nothing further is found wanting, the laboratory issues the compliance report. No compliance report is issued until everything even remotely connected with the product that will be imported complies with the requirements of all agencies: electrical, safety, communications, public telephones, user manual translations, etc.

Getting a compliance report does not mean the product is homologated; only that it meets the requirements for homologation. The report now goes to the proper ministry (or ministries), to fulfill the legalities of filing and posting. Six to eight additional weeks may pass from the time products are fully technically compliant and the report has been sent, before all the various ministries issue their own compliance reports to the PTT. Then it, in turn, produces the certification of compliance. It is now that the coveted approval number and import permit are granted.

Each complete homologation process can average from six months to a year, and cost anywhere from $50,000 to $100,000. In the end, it is anticlimactic to realize that after all this effort and expense you are authorized to sell in only that one country.

Why Do So Few Pursue Homologation?

It is logical to wonder whether, if homologation is of such value to products like CBF boards, why all companies do not pursue it with determination.

If a company is to seriously attempt wide scale homologation, it will soon learn that there is more to it than just determining what a particular country wants and then tinker it into the product as a kind of design afterthought. In our case, we made the decision when we first started that our software had to allow, among other things, the parameterizing of timings. The hardware is designed from the beginning enable circuits to accept modular components.

This is key to success in homologations. Our knowledge of the PTTs, acquired during those early attempts to get international approvals taught us the differences that must be watched for to successfully procure homologation. We realized that these differences lie principally in the signaling area what goes on to the telephone network in the timings of that signaling, and in the development of cadence monitors for it.

A CBF company expecting to sell its products internationally cannot get by with having one software that does do all it must to be fully parameterizable it for ringer differences, for how long the boards have to signal the specific telephone network. Every analog and digital circuit having anything to do with telephony must be fully programmable at the software level.

When someone looks at a CBF board designed with the international market in mind, one of the first things that is noticed, particularly in the DAA area, is empty slots, available for different modules. For a board to be sufficiently programmable it must allow the easy insertion of components for different countries, the adjustment of ring levels, and other parameters. This setup allows the unique country codes defined by the ITU-T to be used as a single trigger. This, as well as other capabilities guarantee customers that when they enter that number into the software, it will load the correct table for the country. This, in turn, programs the CBF card for correct operation with that particular nation's phone lines, as well as for meeting the various legal specifications that are required by each government. Developers using products designed with this kind of foresight are not burdened with these considerations it is all on the board and software, ready to respond transparently. There is no need to worry about things like unmatched signal levels, ringer differences, violation of access rules, and disruption of service, as can be the case with "grey market" products; that is, unapproved boards manufactured elsewhere that are brought into the country and used without knowledge or permission of the authorities.

Homologation and Users

Homologation directly affects a system integrator's ability to market products beyond a country's borders. To afford developers the widest global opportunity available, however, a manufacturer must not only be able to guarantee that the product will communicate internationally with any fax machine anywhere on the planet, but that it will give the added value of meeting all of a specific country's official technical requirements. For example, in France if a fax call is attempted to a phone number that is not that of a fax machine, by law the caller may not try again before a specified amount of time. To comply with the law, the system must automatically prevent the user from trying to send to that number again before the requisite amount of time established and required by the French PTT has passed. Other, or grey market, CBF boards continue retrying. The result is that the system integrators and businesses using them find themselves in violation of the law of the republic. This situation can result in stiff fines and even prevent a system integrator from ever importing into that country again. The user may be prohibited from utilizing the equipment again or, in some cases, the equipment itself may be confiscated.

How these often subtle different PTT requirements result in the handling of the transaction should be well and clearly outlined in the CBF boards basic documentation. These are all factors of primary importance. The product should insulate developers and users everywhere from worrying about these matters. The interface to the fax subsystem should bridge all the pitfalls involved in homologation and protocol questions, and do it automatically and transparently. Ideally, customers who develop systems on a manufacturer's Japanese CBF board, for example, should be able to market their products in the United States or Iceland knowing that the working of their software need not change and that all national requirements—technical and regulatory—are being fully complied with.

Other benefits of homologation, particularly for multinational corporate customers may not be as immediately obvious. By standardizing on systems using products that have a wide range of homologations, corporate leaders can make centralized MIS decisions. This enables them to cut costs secure in the knowledge that the deployment of these systems in the countries they operate in will not be obstructed due to technical or regulatory shortcomings. Obviously, no one is homologated in every country on the globe; however, consideration should be given to whether a specific product is present in the world's largest markets, where most corporations have significant offices and fax distribution needs.

Chapter 38

Facsimile Protocol Testing

The Genoa Technology Approach

Overview

Before we looked at facsimile testing we did our homework in trying to identify a need. The feedback we received from the industry was that testing was not needed. The way the industry did it's testing was not very sophisticated. Everybody had a list of phone numbers of different facsimile devices that they could called. If the call went trough everything was OK. If no there was a problem. Believe it or not but nobody felt that there is anything wrong with that. Yes, interoperability problem existed but this was normal. T.30 is a recommendation not a standard. You can interpret it as you wish.

Things are different now. The facsimile industry benefits from the presence of powerful test systems. Facsimile is expanding from an image exchange technology to a unified message exchange technology. Multiple attempts are in development to define a unified message exchange format for fax that can be used by other technologies. The goal is to keep the main feature that made facsimile successful and this is the big green button, in other words ease of use. 1997 and 1998 will be the year of the facsimile revolution. At the same time the complexity of the technology is increasing and the time to market is getting shorter.

Facsimile is quickly gaining new dimensions. In 1990, implementing a fax machine meant:

- A V.29 modem
- T.30 state machine
- Data compression: MH/MR/MMR
- Primitive scanner
- Primitive print engine
- Standards defined by V.27ter, V.29, T.30, T.4, T.6

The picture in 1997 is quite different:

- V.29/V.17/V.34 modems
- Good B&W and color scanner
- Good B&W and color print engine
- T.30 state machine for traditional fax
- New T.30 state machine for V.34 with V.8 support
- Data presentation schemes:
  1. MH/MR/MMR compressing schemes
  2. TIFF for Internet Fax
  3. JPEG coding with T.42 color space
  4. BFT file format
- Standards defined by V.27ter, V.29, V.17, V.34, V.8, T.30, T.4, T.6, JPEG, T.42, T.434, IETF TIFF F

Another major achievement is that for the first time the ITU-T recommendations are ahead of the industry. Manufacturers do not need to go through nonstandard capabilities to use the latest in technology while waiting for the standard bodies to catch up.

The goal of testing is to simplify the task of bringing a fax product to market. This means helping manufacturers lower the time to market and increasing the interoperability chances.

The following is the test methodology we defined to achieve this goal.

Testing Methodology

This section describes the test methodology chosen by Genoa technology to test facsimile devices.

There are many approaches to testing facsimile. The most important ones are:

- Interoperability Testing
- Conformance Testing
- Worst Case Testing
- Load Testing
- Burst Testing
- Acceptance Testing
- Benchmark Testing

Our customer's focus is on interoperability and conformnance.

Interoperability

Interoperability is the only real issue. All testing is done to achieve interoperability.

Definition: We define interoperability testing is the process that verifies that the Facsimile Device Under Test (FDUT) interoperates with all existing and future devices.

Interoperability testing is a non-deterministic problem. The main reasons are that we do not have access to future non-released products and that it is practically impossible to recreate all the possible scenarios by using only real facsimile devices. On the other hand the short time to market does not support the complexity of the logistics and effort needed to do a comprehensive job.

Conformance

Definition: Conformance testing is the process that verifies that the FUT obeys the rules defined by the specification it implements.

Unfortunately conformance to a specification does not guarantee interoperability. It only increases the chances of interoperating with devices that implement the same specification. The T.30 recommendation allows scenarios in which two devices that conform to the specification will not interoperate. A typical case occurs when after the CED/DIS arrives from the answerer, the originator waits before sending DCS/TCF a time equal to the answerer timeout for repeating the DIS signal. In this case the DCS/TCFsignals will collide with the DIS signal coming from the answerer. This scenario can be repeated until the stations disconnect.

Conformance testing is a deterministic problem. Tests for each requirement in the specification can be defined. This does not mean the tests can be implemented. Tests that require special actions from the facsimile under test that are outside operator control cannot be implemented.

Controlling the coverage of the conformance testing process allows you to control the testing time. Conformance testing can be automated easier than interoperability testing.

Facsimile Testing in the Real World

The best compromise is to test for conformance during the beginning of the development process and with a selected set of real world devices for interoperability close to the end to test. The selection of the real world devices is made depending on market penetration.

Test Tools

Genoa Technology, Inc. developed its test tools Test Tools based on the above methodology. For conformance we developed test suites and traffic generators. For interoperability testing we developed protocol analyzers and device emulators.

Test Tools for Conformance

Facsimile Test Suites

Test Suites are sets of predefined tests. Each test verifies one conformance requirement. Tests have clear PASS/FAIL criteria.

The way test suites are defined is very simple. Let us consider all the possible scenarios that can be defined between two facsimile devices. Please note that two scenarios that are identical in everything but one parameter value, CFR silence time for example are considered different. Let us call this set the Generic Test Suite (GTS). The generic test suite is an infinite set. The infinite comes from the fact that parameter ranges are contiguous and from the fact that you can theoretically have an infinite number of pages to send or you can loop indefinitely. A typical scenario for indefinite looping in T.30 is the retransmission at RTN. T.30 does not specify a maximum number of retransmissions so you can define tests for any number of them.

A Real Conformance Test Suite is a finite subset of the GTS. This is obtained by reducing the infinite to finite by changing contiguous ranges into discrete ones (min., max. and step), limiting the number of pages to be sent and limiting the number of times the tests can go through a loop.

Coverage is a term that describes how well a real test suite maps over the generic test suite. Test suites that offer good coverage give better interoperability chances than the ones that offer a lower coverage.

To build a real test suite that is useful, economical filters need o be applied.The intent of the economical filters is to minimize the testing cost and testing time. Applying an economical filter means focusing on what is relevant for a desired goal. You can generate comprehensive tests (long testing time and cost but good coverage), functional tests (poor coverage, short testing time and cost), worst case tests (focus on extreme cases and known anomalies), etc.

The facsimile industry has defined multiple sets of test suites. Test Suite Specifications available now are:

USA)	TSB 85 - TIA/EIA-465-A and YTIA/EIA-466-A Conformity Test Standard
Germany	DTS - Der Telefax Standatd (The Telefax Standard)
Europe	ETS 300 242 - Group 3 Fax Terminal Testing and Type Approval
France	NT/SPT/SCE/STD/110 - Description of tests for the Group 3 T.30 Protocol

Genoa Technology, Inc. implements all these test suites plus two of it's own choice. These additional test suites are:

1. T.30 Conformance Test Suite – a very comprehensive test suite with very good coverage
2. Worst Case Test Suite – implements all the T.30 anomalies that we encountered in the real world plus a selected number of tests that stress the facsimile protocol to it's limits.

Facsimile Traffic Generator

The facsimile traffic generator simulates one end of the communication. The traffic generator must be fully configurable. This means the user has control over the execution sequence and the parameter values. A typical facsimile traffic generator will allow the user to build his own call path and modify the value of any parameter (flag sequence length, silence time, etc.).

Genoa Technology, Inc. offers a traffic generator, FaxScript, that consists of a scripting tool specially designed to build facsimile calls. The scripting tool is like an interpreter with procedures, flow control, support for variables and specific facsimile functions like SendTrainingCheckSequence(), ReceiveFacsimileMessage(), etc. FaxScript is designed to implement any T.30 scenario plus any other scenario that uses T.30 like signals (same modulations).

Test Tools for Interoperability

Facsimile Protocol Analyzers

Protocol Analyzers passively capture facsimile traffic and analyzes it according to the requirements defined by the T.30 recommendation.

Facsimile Device Emulator

The Facsimile Device Emulator is that testing device that can emulate the personality of a real world facsimile machine. The emulated behavior is at the level of the protocol only. This means that characteristics like modulation parameters and form factor are not emulated.

How does a facsimile device emulator work? The first thing is to define a model for the device. In the case of facsimile this is a mathematical model for the state machine defined in T.30. This model consists of the set of data parameter values and event transitions that define one state machine. We call the model the personality profile of a facsimile machine.

The next step is to measure real world device behavior and fill in the data needed by the model. This process is called profiling. During profiling the values and event transitions of the personality profile are measured. Statistical measurement is the best known approach for good profiling. This means that all parameters end event transitions are measured multiple times and the statistical data is stored into the profile. The larger the personality profile, the more accurate the emulation is. The larger the profile, the more expensive the profiling process is. One important note is that not all parameters in the personality profile can be automatically measured. Parameters like Off Hook to CNG delay for the originator are measured using a digital oscilloscope. These first two steps happen at Genoa.

The third step is to generate traffic using the device profile. This is as simple as selecting the desired emulation from a list. In this moment the Facsimile Device Emulator will respond in the same way as the real device to any T.30 input. In other words the facsimile device emulator works in the same way as a facsimile device. The facsimile device emulator and the real device must have identical responses as result of the same stimuli. This means that if you run a facsimile call to the real device and the same identical call to the emulation of the real device, the resulting captured data for the two calls must be the same.

Using a facsimile device emulator does not necessarily eliminate the need to test with real world facsimile machines. It only reduces the number of machines to test with. Please remember that modulation level parameters such as carrier frequency or phase jitter are not emulated. Only T.30 protocol level behavior is emulated. Most modem aspects can be tested with a good phone network emulator.

The big benefits in using a device emulator are:

- Reduces the need for extensive real world testing
- Allows a large part of the testing process to be automated (at a minimum you eliminate having someone sitting at a fax device to feed paper and dial phone numbers)
- You can force the emulated device to go through call paths that are extremely difficult to obtain in the real world (paths including signals like EOM, RTP, RTN, PPR, EOR are just a few examples of this)

Genoa Technology's facsimile device emulator is called FaxLab. A distinction that is often cost is between the emulator and the traffic generator (or simulator). There is a big difference. The difference is the profile. The responses in a traffic generator come from what the user defines or from a predefined test. The responses of an emulator come from an estimation of what the emulated device would answer in the same situation.

How and When to Use The Testing Tools

If you Develop or Integrate T.30

Development

During early development the most important tool is the Traffic Generator. What you need at this stage is to test very specific peaces of code that you develop. The traffic generator is the one that gives you the flexibility to do generate any scenario you need.

End of Development

This means that the code is put together, individual peaces are tested and now it is time for the system to be checked. At this stage the most useful tools are the test suites. Test suites are traffic generators with predefined traffic configurations. You need to run hundreds and hundreds of tests that take your implementation through all possible branches in the code. Predefined test suites are the best tools for this. They spare you the time to think about all possible scenarios you need and the time to implement them using a traffic generator. A good comprehensive conformance test would be the choice. This is strongly recommended especially if the code is new.

Using a test suite that tests worst case situation is also useful. This condenses the set of anomalies observed in the real world.

QA

Here you need most of the test tools. Test suites are used to validate that the code is bug free, device emulator to check as much of the real world interoperability as possible and protocol analyzers for debugging problems that show only with real world devices. Please note that testing with a few selected real world devices is a must.

Test automation is the main concern. Our experience shows that by the time you get to the testing, the project is already delayed and it is out of budget. It is dangerous to cut corners in the amount of testing that needs to be done. Automation will buy some time and money.

Regression Testing

Regression testing is a quick check as opposed to the extensive testing after a full development cycle. The best tool for this stage is either a selected number of test cases from a test suite (the selection criteria depends on the particularities of the device to test) or a selected set of device emulations. We recommend the selected set of device emulations or previously failed tests.

Product Support (Field Engineering/Technical Support)

Facsimile Protocol Analyzers are the best tools technical support and field engineers. The protocol analyzer sits between the two devices that interoperate and records al the traffic exchanged between the two. Portability is a major issue for anybody that does field engineering job. A portable protocol analyzer is a perfect tool for them.

Most of the problems are easy to identify once they are recorded and analyzed. The protocol analyzer's internal analyzing capability should get you what you need. The built in capability to record the eye pattern gives enough data to asses the line quality.

If the problem is more complicated, just mail the captured data to the engineering department where they can take a harder look at it. They can use a traffic generator to load the data and automatically build a script that will simulate that particular scenario.

Multi-Channel Fax Systems

This type of product is the multi-channel version of the product defined in the previous section. In addition to the testing described above, this type of device requires concurrence testing. The purpose of the concurrence testing is to test the ability of the device under test to deal with multiple calls coming in the same time. The only relevant parameters for the facsimile calls are the sequencing and the complexity of the calls.

If your Product Carries T.30 Traffic

This type of product is also known as facsimile demod/remod device. These are actually facsimile gateways. A typical example is a wireless network. The facsimile traffic is demodulated on one end of the wireless network, transmitted over the network and re-modulated at the other end.

The device under test in this case is not one end of the communication channel but the communication channel itself.

In reality these tests implement two T.30 state machines, one at each end. In many cases because of network delay, creative solutions are required. T.30 contains a lot of ways that allow one end to buy time, sending early flags instead of a command is supposed to reset the T2 timer and force the mote machine to wait for another T2 seconds is one of them. Unfortunately strict conformance to T.30 is still an issue and you cannot count on these details to be implemented in all facsimile devices.

Problems may occur because of the T.30 implementations but also because of the protocol use after demodulation to transfer the data.

The best test setting for this type of device consists of two testers that are linked with each other, each sitting on one end of the communication channel. The most useful scenario is to use two device emulators and run pairs of tests simulating various emulations on each side (Canon to

Canon, Canon to Sharp, Panasonic to Sharp, etc.). FaxLab, which is the Facsimile Device Emulator from Genoa Technology, Inc. has the ability to work in such a setting. Testing is fully automated in this case.

For those gateways that use T.30 tricks to buy time, a traffic generator is a very useful tool to build a small set of tests that can qualify facsimile devices that can operate on the network. In this case a quick check of the devices involved in the communication must be made before a decision can be made that there is a problem with the network.

The facsimile protocol analyzer is an indispensable tool for debugging problems in the real world. An additional function of the analyzer is to help isolate problems that are caused by the protocol between the gateways. Having 2 protocol analyzers, one on each end, helps identify the inputs and outputs for each end allowing the test engineer to isolate the problem more easily.

Typical Errors

This section lists a few only of the errors encountered while profiling real world devices.

- Certain short silences cannot be obeyed.
- Transmission of CNG ceases at CED detection.
- Adds echo protect tone on all V.29 modulations.
- Cannot always or never responds to first DIS.
- Cannot always respond to first post-message command.
- CED transmitted at -1 dBm.
- CED transmitted at wrong frequency.
- Command Received exits before 200 ms silence.
- Disconnects after receiving 0 scan lines in ECM.
  Disconnects after receiving 0 scan lines in non-ECM.
  Disconnects after receiving no RTC in ECM.
  Disconnects after receiving no RTC in non-ECM.
- Disconnects after receiving a CRP response to DIS.

- Disconnects after receiving an EOP command.
- Disconnects after receiving an MCF response to EOP.
- Disconnects after receiving any PRI-Q command.
- Disconnects or goes silent inexplicably.
- Goes silent after receiving 0 scan lines in ECM.
  Goes silent after receiving 0 scan lines in non-ECM.
  Goes silent after receiving no RTC in ECM.
  Goes silent after receiving no RTC in non-ECM.
- Ignore PIP or PIN responses as if not received, indefinitely.
- Ignores BAD FCS responses and waits for T4 elapse to retransmit.
- Ignores CRP response and waits for T4 elapse to retransmit.
- Incorrect amount of fill bits according to signaled MSLT.
- Incorrect count of RCP frames.
- Incorrect duration of TCF zero bits.
- Incorrect duration of V.21 leading flags.
- Incorrect EOL count in RTC or EOFB.
- Incorrect FCS in third RCP.
- Incorrect HDLC control octet in third RCP.
- Incorrect MSB in FCF after receiving DTC.
- Incorrect MSB in FCF after transmitting DTC.
- Incorrect PP synchronization sequence length.
- Incorrect silence before transmitting CRP.
- Incorrect T4 delay to repeat unanswered command.
- Marginal or incorrect transmitted TCF length (less than 1350 ms).
- Never responds CRP to FCS error.
- Never responds RTP following bad pages.
  Never responds RTP or RTN following bad pages.
- Not able to retransmit message after receiving RTN.
- Omits echo protect tone on all V.17 modulations.

- Omits echo protect tone on all V.27 modulations.
- Omits echo protect tone on all V.33 modulations.
- Protocol frames containing FCS errors are not rejected.
- Responds DCN or RTN when RTC contains only two EOL codes.
- Responds RTP for MCF after responding RTN once.
- Response frames having FCS errors are detected, but used.
- Sends spurious bits at beginning of message.
- Sends spurious bits at beginning of TCF.
- Sends DCN after receiving a certain percent bad scan lines.
- Sends DCN after receiving 0 scan lines in ECM.
  Sends DCN after receiving 0 scan lines in non-ECM.
  Sends DCN after receiving no rtc in ECM.
  Sends DCN after receiving no rtc in ECM.
- Sends DCN after receiving any CRP response.
- Sends DCN after receiving any PIP or PIN response.
- Sends DCN after receiving CRP response to post-message command.
- Sends DCN after receiving any RTN response.
- Sends DCS containing ECM and non-zero MSLT.
- Sends DCS with fine resolution, exceeding DIS abilities.
- Sends DCS with unlimited length, exceeding DIS abilities.
- Sends DCS with V.17 data rate, exceeding DIS abilities.
- Sends EOR with invalid FCF2 code.
- Sends no flags after third RCP.
- Solitary or variable count of V.21 trailing flags.
- Timer T2 not reset at Command Received <FLAG?> moment.
- After RTC, message carrier remained on four more minutes.
- Brief loss of energy near end of DCN.

- Brief losses of energy during MSG.
  Brief losses of energy during PP.
- Cannot always or never responds to second DCS.
- Disconnects after receiving a CTC before fourth PPR.
- Disconnects after receiving any DTC.
- Disconnects after sending an EOR response.
- Garbles all remaining V.17 Partial Pages after CTC.
- Goes silent after receiving a bad-FCS response.
- Never answers DCS using V.17 7200 bit/s or V.17 9600 bit/s.
- Never answers the first CTC.
- Never sends DCN after the final failed DCS-TCF attempt.
- Omits echo protect tone on V.17 message modulations.
- FaxLab neither profiles nor emulates this anomaly.
- Responds DCN with an incorrect FCF MSB, after many bad EOPs.
- FaxLab neither profiles nor emulates this anomaly.
- Responds FTT to any DCS-TCF using V.33 modulation.
- Retransmits an unsolicited response when next command is late.
- Sends 16 EOR-NULLs before finally disconnecting.
- Sends CFR after receiving 0 scan lines in non-ECM.
- Sends DCN instead of EOR command.
- Sends DIS after receipt of Frame Check Sum error.
- Sends DIS after receipt of scan line errors.
- Sends DIS quickly after third bad-FCS EOP.
- Sends EOR, not CTC, upon the fourth PPR response.
- Sends ERR instead of ECM post-message response.
- etc.

For more information, visit: www.gentech.com

Glossary

Why reinvent the wheel...

This Glossary is a compilation of three sources:

Newton's Telecom Dictionary, by Harry Newton
Published by Telecom Library Inc.
212-691-8215 or 800-LIBRARY
Harry's dictionary is the first book in any CTI/telecom library.

PC Telephony, by Bob Edgar
President, Parity Software Development Corporation
Published by Newton's Flatiron Publishing Inc.
212-691-8215 or 800-LIBRARY
This was the first and continues to be the best book on Voice Processing.

Understanding Fax Technology
GammaTech Publication GTP1001
GammaLink Division of Dialogic Corporation
Santa Clara, CA
408-969-5200

This section, as well as this book, could not have been completed without their assistance.

10 BASE 2
IEEE standard for baseband Ethernet at 10 Mbps over coaxial cable to a maximum distance of 185 meters. Also known as "Thin Ethernet"

10 Base-T
An IEEE standard for operating Ethernet LANs on twisted pair wiring that appears like telephone cabling. Sometimes old cabling will not work.

A4
Basic Group 3 standard defined for the scanning and printing of a page 215 mm (8.5 in) wide. An A5 page is 151 mm (5.9 in) wide, and the A6 is 107 mm (4.2 in) wide.

ACTIVITY REPORT
Provides a record of transmission time, date, size of the file, recipients telephone number, transmission success or failure, the senders name, and other pertinent information. This is a valuable management tool to get an overview of a company's fax traffic and costs.

AEB
Analog Expansion Bus. Dialogic's name for the analog electrical connection between its network interface modules and its analog resource modules. This is an open technical specification, and individuals can make their own resource modules and/or network interface modules. See also PEB, which is the more modern digital PCM expansion bus.

ANALOG
Comes from the word "analogous," which means "similar to." In telephone transmission, the signal being transmitted C voice, video, or image C is "analogous" to the original signal. In other words, if you speak into a microphone and see your voice on an oscilloscope and you take the same voice as it is transmitted on

the phone line and ran that signal into the oscilloscope, the two signals would look essentially the same. The only difference is that the electrically transmitted signal (the one over the phone line) is at a higher frequency. In correct English usage, "analog" is meaningless as a word by itself. But in telecommunications, analog means telephone transmission and/or switching which is not digital. See ANALOG TRANSMISSION.

ANALOG EXPANSION BUS

See AEB.

ANALOG / DIGITAL CONVERTER

An A/D Converter. Pronounced: "A to D Converter." A device which converts an analog signal to a digital signal.

ANALOG TRANSMISSION

A way of sending signals C voice, video, data C in which the transmitted signal is analogous to the original signal. In other words, if you spoke into a microphone and saw your voice on an oscilloscope and you took the same voice as it was transmitted on the phone line and threw that signal onto the oscilloscope, the two signals would look essentially the same. The only difference would be that the electrically transmitted signal would be at a higher frequency.

ANI

Automatic Number Identification. A phone call arrives at your home or office. At the front of the phone call is a series of digits which tell you, the phone number of the phone calling you. These digits may arrive in analog or digital form. They may arrive as touchtone digits inside the phone call or in a digital form on the same circuit or on a separate circuit. You will need some equipment to decipher the digits AND do "something" with them. That "something" might be throwing them into a database and bringing your customer's record up on a screen in front of your telephone agent as he answers the call. "Good morning, Mr. Smith."

ANSI

American National Standards Institute. A non-government standard-setting organization which develops and publishes standards for use in the U.S.

API

Application Program Interface. A set of standard software interrupts, calls, and data formats that computer application programs use to initiate contact with network services, mainframe communications programs, or other program-to-program communications. APIs typically make it easier for software developers to create the links an application needs to communicate with the operating system or network.

ASCII

American Standard Code for Information Interchange. A character set which gives a numerical value from 32 to 127 to commonly used letters, numbers and symbols. For example, an upper-case A is assigned the value 64. The IBM PC extended this standard to 255 characters which included symbols required in countries other than the USA such as accented letters. This extended character set is sometimes called the 8-bit ASCII character set to distinguish it from the original standard, which is then called the 7-bit character set since values from 0 to 127 can be represented using 7 bits, and can therefore be transmitted over modem lines using 7 rather than 8 data bits, for example. Computer files stored using these characters are called ASCII text, or sometimes just text files. These files almost always use one byte (an 8-bit binary unit which can take a value from 00000000 = decimal zero to 11111111 = decimal 255 when interpreted as a number) for each character. ASCII files are simpler than word processing files, which have complex codes embedded within them, and have the advantage that most programs can read and write in this format. However, there are only ASCII codes for the most rudimentary formatting information: tab, end of line and end of page markers. All other information such as font size, tab positions and so on is lost when a file is stored as ASCII.

ASCII CHARACTER SET

A character set consisting only of the characters included in the original 128-character ASCII standard. Every character handled by a computer has a number in the ASCII character set. Some computers use an additional 128 ASCII characters beyond the original 128, many of them dedicated to graphics. This is what is referred to by the term "extended character set."

ASYNCHRONOUS COMMUNICATION

A method of data communication in which the transmission of bits of data is not synchronized by a clock signal but is accomplished by sending the bits one after another, with a start bit and a stop bit to mark the beginning and end of the data unit. Two communicating devices must be set to the same speed, or "baud rate." Asynchronous communication normally is used for transmission speeds under 19,200 baud. Because of the lower communication speeds, normal telephone lines can be used for asynchronous communication.

ATM

Asynchronous Transfer Mode. Also known as "BISDN" and cell relay".

AUTOMATIC COVER LETTER

Feature that allows the user to automatically attach a cover letter to the document being faxed. This is especially convenient when sending material such as spreadsheets, for example.

AUTOMATIC REDIAL

Provides for the automatic redialing of a fax number in the event the receiving line is busy or an error occurred in faxing the document. Some products allow the user to specify the redial attempts and correct specific errors to be used when redialing.

AUTOMATIC ROUTING

Allows incoming faxed documents to be automatically routed to the addressed individual on a LAN or centralized system. Currently, technology provides several methods of accomplishing this using DID, OCR or DTMF techniques.

APPLICATION PROGRAMMING GENERATOR (API)

A program to generate actual programming code. An applications generator will let you produce software quickly, but it will not allow you the flexibility had you programmed it from scratch. Voice processing "application generators", despite the name, often do not generate programming code. Instead they are self-contained environments which allow a user to define and execute applications.

AUDIO MENU

Options spoken by a voice processing system. The user can choose what he wants done by simply choosing a menu option by hitting a touchtone on his phone or speaking a word or two. Computer or voice processing software can be organized in two basic ways C menu-driven and non-menu driven. Menu-driven programs are easier for users to use, but they can only present as many options as can be reasonably spoken in a few seconds. Audio menus are typically played to callers in automated attendant/voice messaging, voice response and transaction processing applications. See also MENU and PROMPTS.

AUDIOTEX

A generic term for interactive voice response equipment and services. Audiotex is to voice what on line data processing is to data terminals. The idea is you call a phone number. A machine answers, presenting you with several options, "Push 1 for information on Plays, Push 2 for information on movies, Push 3 for information on Museums." If you push 2, the machine may come back, "Push 1 for movies on the south side of town, Push 2 for movies on the north side of town, etc." See also INFORMATION CENTER MAILBOX.

AUTOMATED ATTENDANT

A device which is connected to a PBX. When a call comes in, this device answers it and says something like, "Thanks for calling the ABC Company. If you know the extension number you'd like, enter that extension now and you'll be transferred. If you don't know it, enter "0" (zero) and the live operator will come on. Or, wait a few seconds and the operator will come on anyway." Sometimes the automated attendant might give you other options, such as, "dial 3" for a directory. Automated attendants are very new. They are connected also to voice mail systems ("I'm not here. Leave a message for me."). Some people react well to automated attendants. Others don't. A good rule to remember is before you spring an automated attendant on your people/customers/subscribers, etc., let them know. Train them a little. Ease them into it. They'll probably react more favorably than if it comes as a complete surprise. The first impression is rarely forgotten, so try to make it a good experience for the caller. See also DIAL BY NAME.

BACKBONE
The part of a communications network that carries network traffic between access devices.

BAUD
The number of changes in signal state per second in a signal sent by a modem A baud may contain four or more bits. Sometimes confused with BPS, the bits per second transmitted on the channel.

BAUD RATE
The transmission rate of a communications channel. Technically, baud rate refers to the maximum number of changes that can occur per second in the electrical state of a communications circuit. Under the RS-232C communications protocol, 300 baud is likely to equal 300 bps, but at higher baud rates the number of bits per second transmitted is actually higher than the baud rate because one change can represent more than one bit of data. For example, 1,200 bps is usually sent at 600 baud by sending two bits of information with each change in the electrical state of the circuit.

BINARY
A term used to refer to a system of numbers having 2 as its base, such as the digits 0 and 1. A binary system of numbers is used by digital computers because they can only represent data as two states: on or off. (See Digital.)

BISDN
See ISDN.

BIT
Binary digit. The smallest amount of information in a binary system, a 0 or 1 condition.

BPS
Bits per second.

BUFFER
A defined amount of memory assigned to holding data temporarily. This type of information storage is often used to compensate for differences in processing speeds between computers, components, and peripherals, or to free the computer to carry out other operations as when, for example, a complete file is downloaded to a printer's buffer.

BYTE
A group of 8 bits, making up a single memory location. Most computers cannot address a bit, they can only address byte.

BROADCASTING (FAX)
This procedure allows the user to fax a document to a group of people or companies and if desired, personalize each document. Groups can be temporarily or permanently stored in the telephone directories for repeated broadcasting.

BUSY
In use. "Off-hook". There are slow busies and fast busies. Slow busies are when the phone at the other end is busy or off-hook. They happen 60 times a minute. Fast busies (120 times a minute) occur when the network is congested with too many calls. Your distant party may or may not be busy, but you'll never know because you never got that far.

CALL CENTER
A place where call are answered and calls are made. A call center will typically have lots of people (also called agents), an auto-

matic call distributor, a computer for order-entry and look-up on customers' orders. A Call Center could also have a predictive dialer for making lots of calls quickly.

CALL COMPLETION

This is industry jargon for "putting the call through". When a call has been completed, there is an unbroken ("complete") circuit made between the caller and recipient of the call. This circuit is known as the talk path.

CALL PROGRESS ANALYSIS

The automated determination by a piece of telecommunications equipment as to the result of dialing a number. For example, the result of the analysis might be a busy tone, ringing at the other end but no answer after a pre-set number of rings, an answered call and so on. The analysis involves detecting the various call progress tones which will be generated by the telephone network as the call is put through.

CALL PROGRESS MONITORING

Closely analogous to call progress analysis, call progress monitoring may be active during the entire length of a conversation. For example, when a call is placed across a PBX or in a country which does not provide for loop current drop disconnect supervision, it may be necessary for equipment to monitor for a "re-order" or dial tone to determine that the caller hung up. This would be classified as call progress monitoring since it must take place during the entire call, not just when a number is dialled or a transfer is initiated.

CALL PROGRESS TONE

A tone sent from the telephone switch to tell the caller of the progress of the call. Examples of the common ones are dial tone, busy tone, ringback tone, error tone, re-order, etc. Some phone systems provide additional tones, such as confirmation, splash tone, or a reminder tone to indicate that a feature is in use, such as confirmation, hold reminder, hold, intercept tones.

CALLER ID

A name for a service which displays the calling party's telephone number on a special display device.

CALLING TONE

See CNG.

CARRIER FREQUENCY

The frequency of a carrier wave.

CARRIER WAVE

A wave having at least one characteristic that may be varied from a known reference value by modulation.

CAS

Communicating Applications Specification. A high-level API (application programming interface) developed by Intel and DCA that was introduced in 1988. CAS enables software developers to integrate fax capability and high-speed, error-corrected file transfer into their applications.

CBF

Computer-Based Fax.

CCITT

Consultative Committee for International Telephone and Telegraph.

On March 1, 1993, the CCITT changed its name to the International Telecommunications Union (ITU-T).

CED

Called Station Identification. A tone used in the hand-shaking used to set up a fax call: the response from a fax machine to the called machines CNG tone.

CENTRAL OFFICE

A telephone company facility where subscriber's lines are joined to switching equipment for connecting other subscribers to each other, locally and long distance.

CENTREX

Centrex is a business telephone service offered by a local telephone company from a local central office. Centrex is basically single line telephone service delivered to individual desks (the same as you get at your house) with features, i.e. "bells and whistles," added. Those "bells and whistles" include intercom, call forwarding, call transfer, toll restrict, least cost routing and call hold (on single line phones).

Centrex is known by many names among operating phone companies, including Centron and Cenpac. Centrex comes in two variations C CO and CU. CO means the Centrex service is provided by the Central Office. CU means the central office is on the customer's premises.

CFR

Confirmation to Receive frame.

CHANNEL

A path of communication, either electrical or electromagnetic, between two or more points. Also called a circuit, facility, line, link, or path. Typically, what a subscriber rents from the telephone company.

CIG

Calling Subscriber Identification. A frame that gives the caller's telephone number.

CLASS 1

The Class 1 interface is an extension of the EIA/TIA (Electronic Industry Association and the Telecommunications Industry Association) specification for Group 3 fax communication. Class 1 is a series of Hayes AT commands that can be used by software to control fax boards. In Class 1, both the T.30 (the data packet creation and decision-making necessary for call setup) and ECM/BFT (error-correction mode/binary file transfer) are done by the host computer.

CLASS 2

A specification that allows the modem to handle these (Class 1)

T.30 functions in hardware. Class 2.0 is a specification that allows the serial modem to handle T.30 functions in hardware, as well as providing ECM.

CLASS 3

Provides the same specifications as Class 2, as well the parameters for the conversion of image file data into ITU-T T.4 compressed image for transmission, and reversion of the conversion on reception.

CLIENT

The requesting program in a distributed computing system. The "Client" send requests to servers across a network and waits for indication from the server that the request is complete.

CNG

Calling Tone. The piercing "whistle" tone (1,100 Hz) of a fax machine to inform the caller that it is ready to receive a transmission.

COMPRESSION

Changing the storage or transmission scheme for information so that less space (fewer bits) are required to represent the same information. Compressing data means that less space is required for storage and less time for the transmission of the same amount of data. Comes in two flavors: lossless compression, where the original information can be reconstructed precisely, and lossy compression, where something close to the original can be reconstructed but some details may differ.

COMPRESSION ALGORITHM

The arithmetic formulae that convert a signal into a smaller bandwidth or fewer bit.

COVER PAGE

The first page of a fax message. It generally includes a header, typically the sender company's logo, the recipient's name and fax telephone number, the sender's fax and voice telephone numbers, the system's date and time, a message, and a footer.

CRP
Command Repeat.

CSI (CSID)
Called Subscriber Identification. An identifier whose coding format contains a number, usually the telephone number from the remote terminal used in fax.

CONNECTED SPEECH
A technical term used to describe speech made of a series of utterances which come in relatively quick succession without co-articulation. See CO-ARTICULATION. Connected speech is intermediate between discrete speech and continuous speech. Usually applied to the capability of a voice recognizer to recognize words from this type of speech.

CONTINUOUS SPEECH
A technical term used to describe speech made of a series of utterances which come in relatively quick succession with co-articulation. See CO-ARTICULATION. Usually applied to the capability of a voice recognizer to recognize words from this type of speech.

CSI
Called Subscriber Identifier. The "name" of a fax device, transmitted to the fax device at the other end in the course of establishing a fax call. Typically a telephone number and/or company name.

CTI
Computer Telephone Integration. A much better name than "voice processing" for voice processing technology.

DAA
Data Access Arrangement. A device required to hook up Customer Provided Equipment (CPE), usually modem and other data equipment, to the telephone network.

DCN

Disconnect frame. Indicates the fax call is done. The sender transmits it before hanging up; it does not wait for a response.

DCS

Digital Command Signal. Signal sent when the caller is transmitting, which tells the answerer how to receive the fax. Modem speed, image width, image encoding, and page length are all included in this frame.

DECADIC SIGNALING

A fancy way of referring to pulse dialing.

DIAL TONE

The sound you hear when you pick up a telephone. Dial tone is a signal (350 + 440 Hz) from your local telephone company that it is alive and ready to receive the number you dial. If you have a PBX, dial tone will typically be provided by the PBX. Dial tone does not come from God or the telephone instrument on your desk. It comes from the switch to which your phone is connected to.

DIALOGIC

Dialogic Corp, Parsippany, NJ, is one of the leading manufacturers of interactive voice processing equipment and software. They sell equipment through value added resellers, dealers and distributors. Many of their dealers "add value" to the Dialogic components by doing their own specialized software programming, tailoring Dialogic products to particular specialized (and useful) applications.

DID

Direct Inward Dialing. You can dial inside a company directly without going through the attendant. This feature used to be an exclusive feature of Centrex but it can now be provided by virtually all modern PBXs and some modern hybrids. Sometimes spelled DDI, especially in the UK.

DID TRUNKS
Are employed to reduce the number of channels between the PBX and the telephone company central office. DID trunks are one-way trunks. A PBX perceives the DID trunk as one of its single-line telephones and can interpret four-digit dialing.

DIGITAL
A term used to refer to discrete, uniform signals of any kind, not necessarily binary, that do not vary in a continuous manner, as do analog signals. Digital signals are identified by specific values such as on and off, and change instantaneously from one state to another. (See, Binary.)

DIGITAL FACSIMILE
A form of fax in which densities of the original are sampled and quantized as a digital signal for processing, transmission, or storage.

DIGITAL SIGNAL PROCESSOR
A specialized digital microprocessor that performs calculations on digitized signals that were originally analog (e.g., voice) and then sends the results on. Their advantage lies in the programmability of digital microprocessors. DSPs can be used for compression of voice signals to as few as 4,800 bps. DSPs are an integral part of all voice processing systems and facsimile devices.

DIGITAL TRANSMISSION
The use of a binary code to represent information. Analog signals like voice or data, are encoded digitally by sampling the signal many times a second and assigning a number to each sampling. Unlike an analog signal which picks up noise along the way, a digital signal can be reproduced precisely.

DIS
Digital Identification Signal.

DNIS
Dialed Number Identification Service. DNIS is a feature of 800

and 900 lines. Let's say you subscribe to several 800 numbers. You use one line for testing your advertisements on TV stations in Phoenix; another line for testing your advertisements on TV stations in Chicago; and yet another for Milwaukee. Now you get an automatic call distributor and you terminate all the lines in one group on your ACD. You do that because it's cheaper to man and run one group of incoming lines. One queue is more efficient than several small ones, etc. You have all your people answering all the calls. You now need to know which calls are coming from where. So your long distance carrier sends you the call's DNIS C the numbers the person dialed to reach you. Those DNIS digits might come to you in many ways, depending on the technical arrangement you have with your long distance company. In-band or out-of-band. ISDN or data channel, etc. Make sure you understand the difference between DNIS and ANI. DNIS tells you the number your caller called. ANI is the number your caller called from.

DPI

Dots Per Inch. A measure of output device resolution. The number of dots a printer can place in a horizontal inch. The higher the number, the sharper the resolution.

DTMF

Dual Tone Multi-Frequency. A fancy term describing push button or Touchtone dialing. (Touchtone is a registered trademark of AT&T.) In DTMF, when you touch a button on a pushbutton pad, it makes a tone, actually a combination of two tones, one high frequency and one low frequency. Thus the name Dual Tone Multi Frequency. In U.S. telephony, there are actually two types of "tone" signaling, one used on normal business or home pushbutton/touchtone phones, and one used for signaling within the telephone network itself. When you go into a central office, look for the testboard. There you'll see what looks like a standard touchtone pad. Next to the pad there will be a small toggle switch that allows you to choose the sounds the touchtone pad will make C either normal touchtone dialing (DTMF) or the network version (MF).

The eight possible tones that comprise the DTMF signaling system were specially selected to easily pass through the telephone network without attenuation and with minimum interaction with each other. Since these tones fall within the frequency range of the human voice, additional considerations were added to prevent the human voice from inadvertently imitating or "falsing" DTMF signaling digits. One way this was done to break the tones into two groups, a high frequency group and a low frequency group. A valid DTMF tone has only one tone in each group. Here is a table of the DTMF digits with their respective frequencies. One Hertz (abbreviated Hz.) is one cycle per second of frequency.

DTC

Digital Transmit Command.

E-1

The European term for T1. The E1 line bit rate is usually 2.048 Mbps (T1 in the U.S. and Canada is 1.544 Mbps), but variations between the two are not so great that a multiplexer cannot convert between them. Conversion of E1 to T1 involves both the compression law and signaling format.

ECM

Error Correction Mode. When a fax signal is distorted by a noise pulse induced by electrical interference, or any other reason, errors can occur in the bits transmitted over the telephone line. Without ECM, these errors accumulate and may cause the receiving fax device to disconnect, requiring that the call process be restarted from the beginning, regardless of how much correct material has already been received. ECM provides encapsulated data within HDLC frames, giving the receiver an opportunity to check for, and request retransmission of garbled data.

EDI

Electronic Data Interchange. A series of standards providing automated computer-to-computer exchange of business documents (structured business data, editable documents, or electronic transactions such as invoices, purchase order, etc.) between different companies and computers over telephone lines.

EMAIL

Electronic Mail. A popular application on both LANs and WANs which provides communication among users. There is a variety of systems which vary considerably in their level of sophistication. E-mail services can include simple message handling as well as complex file sharing.

EMail GATEWAYS

All E-mail users can send and receive faxes from within the company's E-mail package. Because the fax messaging is integrated directly into the mail system, fax users have virtually no new software to learn and can take advantage of E-mail features such as workgroup distribution lists, attaching graphics files to text messages, support for different network operating systems, and instant notification of received messages. With an E-mail fax gateway, it is possible to send the same message to both mail users and fax recipients.

ELECTRONIC MAIL INTEGRATION

Integrating computer fax and E-mail technology allows the user to send and receive fax documents using a company's E-mail program. The most popular method uses MHS for Novell networks.

EOM

End Of Message frame. A frame from the sender indicating that the message is done, and that Phase B can be repeated. See, EOP.)

EOP

End of Procedure frame. A frame indicating that the sender wants to end the call.

ETHERNET

A LAN used for connecting computers, printers, workstations, terminals, etc., within the same building. Ethernet operates over twisted pair wire and over coaxial cable at speeds up to 10 Mbps.

FACSIMILE

Facsimile or fax equipment allows information (written, typed, or graphic) to be transmitted through the switched telephone system and printed at the other end. The sending fax scans the material to be sent, digitizing it into binary bits and sending those through a modem to the receiving fax, which essentially reverses the process and outputs it through a printer.

FAX

Abbreviation for facsimile.

FAX ACTIVITY RECORDS

Detailed activity logs that provide the status of tasks: the pending log for tasks still being processed, the received log for incoming faxes; and the sent log. for outgoing faxes.

FAXBIOS

An "API" used for in-application faxing. Developed by WordPerfect Corp. and Everex Systems, Inc.

FAX BOARD

A specialized synchronous modem designed to transmit and receive facsimile documents. Many also allow for binary synchronous file transfer and V.22 communication.

FAX SERVER

In a LAN, a PC or a self-contained unit that has fax circuitry accessible to all the network's workstations. The server receives requests for fax services and manages them so that they are answered in an orderly, sequential manner. It is used to send and receive faxes by any network user, sharing the common resource of one or more fax boards. Depending on application, a fax server may have a specialized interactive voice response system that routes faxes to a fax machine the user designates by touchtone numbers. The receiving unit may be the user's or one designated by him.

FAX BROADCAST

See Broadcast Fax.

FAX MAIL

Analogous to, and perhaps a feature of, voice mail. Fax mail allows a caller to fax a message rather than speaking a message. Fax messages may be retrieved by the mailbox owner from a fax machine or desktop PC which is able to access the stored file and display it as an image on the computer screen. Some voice mail systems allow fax messages to be incorporated into mailboxes.

FOD

Fax-On-Demand. See Fax-On-Demand.

FAX-ON-DEMAND (FOD)

An enhanced fax technology, advanced Store And Forward. A typical use for fax on demand is to provide product information to potential customers. A caller dials a voice processing unit and selects one or more documents of interest using touchtone menus. If the caller is calling from a fax machine, transmission can being immediately (this is called one-call or same-call faxing). If the caller is using a telephone rather than a fax machine, a fax number can be entered in response to a menu prompt and the fax on demand system will make a later call to that number to deliver the document.

FAX STORE AND FORWARD

This refers to the ability of a computer to store a received fax document as a file stored on a hard drive and re-transmit the document in a subsequent call. Analogous to voice store and forward, which simply means "record" and "play back".

FAX SYNTHESIS

The ability of a computer to create a fax document from stored ASCII text, word processing, database, spreadsheet or other information.

FRAME
A group of data bits in a specific format, with a flag at each end to indicate the beginning and end of the frame. The defined format enables network equipment to recognize the meaning and purpose of specific bit.

FRAME RELAY
Frame relay switching is a form of packet switching, but uses smaller packets and requires less error checking than traditional forms of packet switching. Like traditional X.25 packet networks, frame relay networks use bandwidth only when there is traffic to send. Frame relay does not support voice.

FREQUENCY
The number of complete oscillations per second of an electromagnetic wave.

FTT
Failure-to-train signal.

FULL DUPLEX
A communications protocol in which the communications channel can send and receive signals at the same time.

FSK
Frequency Shift Keying. A modulation technique for translating 1s and 0s into something that can be carried over telephone lines, like sounds. A 1 will be assigned a certain frequency of tone, and a 0 another tone. The transmission of the bits keys the sounds to shift from one frequency to the other.

GPI
A GammaFax Programmer's Interface. C-level programming language, Real-time applications for fax switches and gateways. DOS or OS/2 operating systems.

GROUP 1
Analog fax equipment, according to Recommendation T.2 of the ITU-T. It sends a US letter (82 by 11") or A4 page in about six

minutes over a voice-grade telephone line using frequency modulation with 1.3 KHz corresponding to while at 2.1 KHz to black. North American six-minute equipment uses a different modulation scheme, and is therefore not compatible.

GROUP 2

Analog fax equipment, according to Recommendation T.3 of the ITU-S. It sends a page in about three minutes over a voice grade telephone line using 2.1 KHz AM-PM-VSB modulation.

GROUP 3

A digital fax standard that allows high-speed, reliable transmission over voice grade phone lines. All modern fax devices use Group 3, which is based on ITU-S Recommendation T.4. (Most common worldwide, accounts for more than 90% of all fax machines.)

GROUP 4

A ITU-S fax standard primarily designed to work with ISDN. It is considered difficult to implement, and is not in widespread use owing to the low penetration of ISDN (Group 4 cannot work on non-ISDN lines).

(Group 4's future is questionable.)

HALF DUPLEX

A communications protocol in which the communications channel can handle only one signal at a time. The two stations alternate their transmissions.

HANDSHAKING

An exchange of signals between the fax transmitter and the fax receiver to verify that transmission can proceed, determine which specifications will be used, and to verify reception of the documents sent.

HDLC

High-Level Data-Link Control Standard. It always contains a frame called the Digital Identification Signal (DIS), which describes the standard ITU-T features of the machine. It can also contain two other frames: a Non-Standard Facilities (NSF) frame, which tells the caller about vendor-specific features, and, usually, a Called Subscriber Identification (CSI or CSID) frame, which contains the answerer's telephone number.

HOMOLOGATION

The process of obtaining approval from the local regulatory authorities to attach a device to the public telecommunications network. See also PTT.

HOST

The computer in which a fax board or data modem board resides.

HOT KEY

Refers to TSR utilities in DOS or filters in a Windows environment that allow users to fax without leaving their present application. The ability to send a fax from within an application is one of the most important features of computer fax technology.

HUFFMAN ENCODING

A popular lossless data compression algorithm that replaces frequently occurring data strings with shorter codes. Some implementations include tables that predetermine what codes will be generated from a particular string. Other versions of the algorithm build the code table from the data stream during processing. Huffman encoding is often used in image compression. (See, Modified Huffman Code.)

HUNT

Refers to the progress of a call reaching a group of lines. The call will try the first line of the group. If that line is busy, it will try the second line, then it will hunt to the third, etc. See also HUNT GROUP.

HUNT GROUP

A series of telephone lines organized in such a way that if the first line is busy the next line is checked ("hunted") and so on until a free line is found. Often this arrangement is used on a group of incoming lines. Hunt groups may start with one trunk and hunt downwards. They may start randomly and hunt in clockwise circles. They may start randomly and hunt in counter-clockwise circles. Inter-Tel uses the terms "Linear, Distributed and Terminal" to refer to different types of hunt groups. In data communications, a hunt group is a set of links which provides a common resource and which is assigned a single hunt group designation. A user requesting that designation may then be connected to any member of the hunt group. Hunt group members may also receive calls by station address.

ICFA

International Computer Facsimile Association. Formed in 1991, its members include the leading companies in the communications and computer industries.

INFORMATION PROVIDER

A business or person providing information to the public for money. The information is typically selected by the caller through touch tones, delivered using voice processing equipment and transmitted over tariffed phone lines, e.g., 900, 976, 970. Typically, billing for information providers' services is done by a local or long distance phone company. Sometimes the revenues for the service are split by the information provider and the phone company. Sometimes the phone company simply bills a per minute or flat charge.

INTERACTIVE VOICE RESPONSE (IVR)

Think of Interactive Voice Response as a voice computer. Where a computer has a keyboard for entering information, an IVR uses remote touchtone telephones. Where a computer has a screen for showing the results, an IVR uses a digitized synthesized voice to "read" the screen to the distant caller. Whatever a computer can do, an IVR can too, from looking up train timetables to moving calls around an automatic call distributor (ACD). The only limitation on an IVR is that you can't present as many alternatives on a phone as you can on a screen. The caller's brain simply won't remember more than a few. With IVR, you have to present the menus in smaller chunks.

Some people use IVR as a synonym for voice processing, or all computer-telephone integration technology involving spoken responses from the computer.

INTERRUPT

The temporary pause of a task caused by an event outside that process. An interrupt signal from hardware, such as a modem in a PC, temporarily suspends other ongoing tasks while the CPU performs the task requested by the interrupting device. Once the routine is completed, the CPU returns to the original tasks.

ISDN

Integrated Services Digital Network. A collection of standards that define interfaces for, and operation of, digital switching equipment developed by carriers, equipment manufacturers, and international standards organizations. It is intended to form the basis for the next generation telephone network and is currently being implemented by carriers throughout the world. Instead of one analog telephone line, there would be two 64 kbps "bearer" lines and one 16 kbps data line. Each bearer line could carry voice, video, data, images or combinations of these. As the name implies, it would be a point-to-point digital system.

ISO

The International Standards Organization. Organization in Paris, devoted to developing standards for international and national data communications. The U.S. representative to the ISO is ANSI.

ITU-T

International Telecommunications Union-Telecommunications. One of four permanent parts of the International Telecommunications Union, based in Geneva, Switzerland. It issues recommendations for standard applying to modem, packet switched interfaces, V.24 connectors, etc. Although it has no power of enforcement, the standards it recommends are generally accepted and adopted by industry. Until March 1, 1993, the ITU-T was known as the CCITT.

LAN

Local Area Network. A short-distance network, within a building or campus, used to link computers and peripheral devices under a form of standard control.

LOCAL LOOP

The wire which passes between your home phone and the phone company. Generally a length of good, old-fashioned copper wire.

LOOP

1. Typically a complete electrical circuit. 2. The loop is also the pair of wires that winds its way from the central office to the telephone set or system at the customer's office, home or factory, i.e. "premises" in telephones. 3. In computer software. A loop repeats a series of instructions many times until some prestated event has happened or until some test has been passed.

LOOP CURRENT DETECTION

When a fax board (or any other modem or telephone) seizes the line (i.e., completes the connection between tip-and-ring terminals of the telephone cable), current flows from the positive battery supply in the telephone central office, through the twisted pair in the loop, through the board, and back to the central office negative terminal where it is detected, showing that this telephone line is off hook. The fax board also detects the loop current and can detect problems such as disconnects, shutting down the connection or a busy signal, making it wait and redial.

LOSS-LESS COMPRESSION CODING

A coding designed not to lose any data when compressing or restoring an image.

MAILBOX

A set of stored messages belonging to a single owner. Typically, these will be recorded voice messages, but increasingly mailboxes also include E-mail and fax documents.

MAPI

Messaging Application Programming Interface. Developed by Microsoft.

MCF

Message Confirmation Frame. Confirmation by the receiver that it is ready to receive the next page, starting Phase C again.

MINI/MAINFRAME FAX SERVERS

Although many businesses generate and store information on mainframes or minicomputers, most computer-based fax hardware is PC-based. Many mini/mainframe fax servers function in the same way as LAN-fax servers: users can send and receive faxes from their terminals, saving both time and money while they increase the quality of their fax transmissions. By being on the same computer system as an organization's data, mini/mainframe servers can provide an EDI-like component for large, vertical applications such as accounting or purchase order systems, broadening the electronic reach of these programs to any recipient with a fax machine.

MODEM

Acronym for modulator/demodulator. Equipment that converts digital signals to analog signals and vice-versa. Modems are used to send data signals (digital) over the telephone network, which is usually analog. A modem modulates binary signals into tones that can be carried over the telephone network. At the other end, the demodulator part of the modem converts the tones back to binary code.

MODIFIED HUFFMAN CODE (MH)

A one-dimensional data compression technique that compresses data in an horizontal direction only and does not allow transmission of redundant data. Huffman encoding is a lossless data compression algorithm that replaces frequently occurring data strings with shorter codes. Often used in image compression.

MODIFIED READ

Relative Element Address Differentiation code. A two-dimensional compression technique for fax machines that handles the data compression of the vertical line and that concentrates on space between the lines and within given characters.

MODIFIED MODIFIED READ

A two-dimensional coding scheme for Group 4 fax, but now finding use with Group 3 machines.

MPS

Multi-Page Signal. A frame sent if the sender has more pages to transmit.

MVIP

Multi Vendor Integration Protocol. Picture a printed circuit card that fits into an empty slot in a personal computer. The slot carries information to and from the computer. This is called the data bus. Printed circuit cards that do voice processing typically have a second "bus" C the voice bus. That "bus" is actually a ribbon cable which connects one voice processing card to another. The ribbon cable is typically connected to the top of the printed circuit card, while the data bus is at the bottom. As of writing (summer, 1991) there were three "standard" communications buses defined and accepted. Two buses were defined by Dialogic Corporation. They are called The Analog Expansion Bus and the PCM Expansion Bus. The other bus (called MVIP) is from Natural MicroSystems. Both companies have co-opted a number of companies to accept their standard. Both buses can handle voice, data and fax. Here is a write-up on MVIP from Mitel Semiconductor, which has adopted MVIP (as have over 30 other manufacturers):

"The MVIP consists of communications hardware and software that allows printed circuit cards from multiple vendors to exchange information in a standardized digital format. The MVIP bus consists of eight 2 megabyte serial highways and clock signals that are routed from one card to another over a ribbon cable. Each of these highways is partitioned into 32 channels for a total capacity of 256 voice channels on the MVIP bus. These serial link from one card to another. They are electronically compatible with Mitel's ST-BUS specification for inter-chip communications. By letting expansion cards exchange data directly, the MVIP bus opens the PC architecture to voice/data applications that would otherwise overburden the PC processor with data transfers. The MVIP bus is equivalent to an extra backplane that is capable of routing circuit switched data.

"MVIP systems generally have two types of cards; network cards and resource cards. They differ by the switching they provide and in the way they are wired to the bus. Network cards almost always provide more flexible switching and can drive either the input or the output side of the bus, although they usually drive the output side of the bus. Resource cards usually provide very little switching and are only able to drive the input side of the bus. Resource cards usually rely on the network cards to do most or all of the switching on the MVIP bus."

MENU

Options displayed on a computer terminal screen or spoken by a voice processing system. The user can choose what he wants done by simply choosing a menu option C either typing it on the computer keyboard, hitting a touchtone on his phone or speaking a word or two. There are basically two ways of organizing computer or voice processing softwareCmenu-driven and non-menu driven. Menu-driven programs are easier to use but they can only present as many options as can be reasonably crammed on a screen or spoken in a few seconds. Non-menu driven systems may allow more alternatives but are much more complex and frightening. It's the difference between receiving a bland "A" or "C" prompt on the screen C as in MS-DOS and receiving a menu of "Press A if you want Word Processing," "Press B if you want Spread Sheet," etc.

MULTI-TASKING

Doing several different tasks at the same time on one computer. This should not be confused with TASK SWITCHING. In task switching, the computer jumps from one task to another, typically in response to a command from you, the user. For example, in task switching, you might temporarily stop your word processing, jump into your communications package, dial up a database, grab some information, then jump back into your word processor and put that new information into the document you're word processing. However, in true multi-tasking, you could have told your computer to dial the database, grab the information and alert you when it had grabbed the material. At that point you could have included it in your document. But in the meantime, you could have been happily doing word processing.

NAP

Network Applications Platform. (Unisys, Blue Bell, PA, term). A public telephone network service with the ability to send and receive facsimile documents to standard Group 3 fax machines.

NETWORK

An interconnected group of systems. Computer networks connect different and all types of computers and terminals and peripherals, as well as communications systems.

NETWORK FAX SERVER

Allows users to send faxes from their network workstations. Many fax servers allow the faxing of documents from the applications that created them, leaving users free to continue with their work. Incoming faxes can be directed back to your workstation as well. Different fax server products offer different ways of handling incoming faxes, but most use one of the following: the fax can be received by an operator who forwards it to the proper workstation, a fax extension can be assigned to particular workstations for touchtone routing, or there can be a direct fax telephone line accessed through a DID line.

NETWORK INTERFACE MODULE

Electronic circuitry connecting a system (typically a PC) to the telephone network. Network interface modules come in as many versions as there are ways of connecting to the telephone network-from simple loop start telephone lines to complex primary rate interfaces (PRI) on ISDN. Usually, the network interface module slides into one of the expansion slots inside a PC. The board transmits and receives messages from the resource modules providing access to the telephone network.

NETWORK MODULE
Dialogic-speak for a voice processing component which connects to a phone line, whether it is an plain only analog line, digital trunk, DID circuit or whatever.

NODE
A point of connection into a network. In LANs, it is a device on the network. In packet switched networks, it is one of the many packet switches that form the network's backbone.

NOISE
Unwanted electrical signals introduced into telephone lines by circuit components or natural disturbances that tend to degrade the line's performance.

NSC
Non-Standard Facilities Command. A response to the called fax DIS response.

NSF
Non-Standard Facilities frame. Information sent by one fax device to another indicating vendor-specific facilities beyond the standard Group 3 requirements.

NSS
Non-Standard Facilities Setup command, a response to an "NSF" frame.

OCTET
The ITU-T standard term for "byte".

OFF-HOOK
When the handset is lifted from its cradle it's Off-Hook. Lifting the hookswitch alerts the central office that the user wants the phone to do something like dial a call. A dial tone is a sign saying "Give me an order."The term "off-hook" originated when the early handsets were actually suspended from a metal hook on the phone. When the handset is removed from its hook or its cradle (in modern phones), it completes the electrical loop, thus

signaling the central office that it wishes dial tone. Some leased line channels work by lifting the handset, signaling the central office at the other end which rings the phone at the other end. Some phones have autodialers in them. Lifting the phone signals the phone to dial that one number. An example is a phone without a dial at an airport, which automatically dials the local taxi company. All this by simply lifting the handset at one end C going "off-hook."

ON-HOOK

When the phone handset is resting in its cradle. The phone is not connected to any particular line. Only the bell is active, i.e. it will ring if a call comes in. See ON-HOOK DIALING and OFF-HOOK.

ON-HOOK DIALING

Allows a caller to dial a call without lifting his handset. After dialing, the caller can listen to the progress of the call over the phone's built-in speaker. When you hear the called person answer, you can pick up the handset and speak or you can talk hands-free in the direction of your phone, if it's a speakerphone. Critical: Many phones have speakers for hands-free listening. Not all phones with speakers are speakerphones C i.e. have microphones, which allow you to speak, also.

ONE-CALL Fax-On-DEMAND

The caller dials an automated service from a fax machine, selects a document through touchtone and then hits the Start button on the fax machine to send or receive a fax. This is "one-call" or "same-call" fax, as opposed to "two-call" fax where the caller enters a PIN code or fax phone number, the computer then calls back in a later call with the document.

ONE-DIMENSIONAL CODING

A data compression scheme that considers each scan line as being unique, without referencing it to a previous scan line. One-dimensional coding operates horizontally only.

OVERRUN
Loss of data that takes place when the receiving equipment is unable to accept data at the rate it is being transmitted.

OPERATOR INTERCEPT
When an invalid number is dialed or an error condition occurs on the network, an operator intercept may occur. In the US, SIT tones are heard followed by a recorded message explaining the problem.

PACKET
A bundle of data, usually in binary form, organized in a specific way for transmission.

PACKET SWITCHING
Sending data in packets through a network to a remote location. The data are subdivided into individual packets of data, each with a unique identification and individual destination address. This way each packet can take a different route and may arrive in a different order than it was shipped. The packet ID allows the reassembling of data in the proper sequence. This is an efficient way to move digital data. Although it has been used with fax messages, is not yet useful for voice.

PAY-PER-CALL
Some phone calls have an added charge which is levied by the phone company and sent to an information provider. Services such as weather, sports scores, stock prices etc. may be provided. They are generally reached by dialing numbers with a special area code or prefix. In the US, national numbers with the 900 area code or local calls with the 976 prefix have added charges. Similar services are now available in many countries. These are termed "pay-per-call" or "premium rate" services.

PBX
Private Branch eXchange. A private (i.e. you, as against the phone company owns it), branch (meaning it is a small phone company central office), exchange (a central office was originally called a public exchange, or simply an exchange). In other

words, a PBX is a small version of the phone company's larger central switching office. A PBX is also called a Private Automatic Branch Exchange, though that has now become an obsolete term. In the very old days, you called the operator to make an external call. Then later someone made a phone system that you simply dialed nine (or another digit C in Europe it's often zero), got a second dial tone and dialed some more digits to dial out, locally or long distance. So, the early name of Private Branch Exchange (which needed an operator) became Private AUTOMATIC Branch Exchange (which didn't need an operator). Now, all PBXs are automatic. And now they're all called PBXs, except overseas where they still have PBXs that are not automatic.

PCM

Pulse Code Modulation. The most common method of encoding an analog voice signal and encoding it into a digital bit stream. First, the amplitude of the voice conversation is sampled. This is called PAM, pulse amplitude modulation. This PAM sample is then encoded (quantized) into a binary (digital) code. This digital code consists of zeros and ones. The voice signal can then be switched, transmitted, and stored digitally.

PEB

PCM Expansion Bus. Dialogic Corp's name for the digital electrical connection between its network interface modules, voice store-and-forward modules, and resource modules. This bus is now open. Technical specifications are available, thus enabling outsiders to create their own resource modules and/or network interface modules.

PEL

Picture Element. Contains only black and white information, no grey shading.

PHASE A, B, C1, C2, D AND E (See below)

The stages in a fax transmission:

Phase A: Establishment.

Phase B: Pre-message procedure.

Phases C1 and C2: In-message procedure, data transmission.

Phase D: Post-message procedure.

Phase E: Call release.

PHASE A
In a fax device's call process, the call establishment, or when the transmitting and receiving units connect over the telephone line, recognizing one another as fax machines. This is the start of the handshaking procedure.

PHASE B
In a fax device's call process, the pre-message procedure, where the answering machine identifies itself, describing its capabilities in a burst of digital information packed in frames conforming to the HDLC standard.

PHASE C
In a fax device's call process, the fax transmission portion of the operation. This step consists of two parts "C1" and "C2", which take place simultaneously. Phase C1, deals with synchronization, line monitoring, and problem detection. Phase C2 includes data transmission.

PHASE D
In a fax device's call process, this phase begins once a page has been transmitted. Both the sender and receiver revert to using HDLC packets as during Phase B. If the sender has further pages to transmit, it sends an MPS frame, and the receiver replies with an MCF and Phase C recommences for the following page.

PHASE E
In a fax device's call process, the call release portion. The side that transmitted last sends a DCN frame and hangs up without awaiting a response.

PIN
Procedure Interrupt Negative.

PIXEL
Picture Element. The smallest area of an original, sampled and represented by an electrical signal. A pixel has more than two levels of greyscale information.

PIP
Procedure Interrupt Positive.

POLLING
Refers to some form of data or fax network arrangement whereby a central computer or fax machine/board very quickly asks each remote location in turn whether they want to send some information. The purpose is to give each user or remote data terminal an opportunity to transmit and receive information on a circuit or using facilities which are being shared. Polling is typically used in a multipoint or multidrop line. It is done mostly to save money on telephone lines.

PPO-UP
See HOT KEY

PORT
An entrance to or exit from a network, the physical or electrical interface through which one gains access. The interface between a process or a program and a communications or transmission facility. A point in the computer or telephone system where data may be accessed. Peripherals are connected to ports.

PORT DENSITY
Jargon for "the number of ports, ie. phone lines, supported by a system" or "the number of ports per voice processing board". "High density" means that a lot of ports are handled by one or a few boards.

POSTSCRIPT
A page description language for PCs, which is standard for many graphics and desktop publishing applications. It allows the creation of elaborate documents. Other applications such as

spreadsheets, word processors, and databases rarely require this level of sophistication.

POTS

Plain Old Telephone Service. The basic service supplying standard single line telephones, telephone lines and access to the public switched network. Nothing fancy. No added features. Just receive and place calls. Nothing like Call Waiting or Call Forwarding. They are not POTS services. Pronounced POTS, like in pots and pans.

PRI-EOM

Procedure Interrupt-End Of Message.

PRI-EOP

Procedure Interrupt-End Of Procedures.

PRI-MPS

Procedure Interrupt-Multipage Signal.

PRINTER EMULATION

Enables the mimicking of a printer-generated document. This way, the outgoing fax will look as if it has come from the printer attached to a computer. This can include full formatting, as well as letterhead, signature, and different graphic images.

PRIVATE BRANCH EXCHANGE

A business phone system, often abbreviated to PBX or PABX ("A" for Automatic).

PROTOCOL

A specified set of rules, procedures, or conventions, relating to format and timing of data transmission between two devices. A standard procedure that two data devices must accept and use to be able to understand each other.

PSK

Phase Shift Keying. A method of modulating the phase of a signal to carry information.

PSTN
See PUBLIC SWITCHED TELEPHONE NETWORK.

PTT
The Post Telephone and Telegraph (PTT) administrations, usually controlled by their governments, provide telephone and telecommunications services in most countries where these services are not privately owned. In CCITT documents, these are the entities referred to as "Operating Administrations."

It is not a simple thing to obtain approval from the PTTs to sell and use telecommunication equipment of any kind in their countries. The world is far from being one in the field of telecommunications. Meeting international requirements typically means providing hardware and software modifications to the product, unique to each country, and then going through an extremely rigorous approval process that can average between six to nine months. Products are required to meet both safety and compatibility requirements.

PUBLIC SWITCHED TELEPHONE NETWORK
Usually refers to the worldwide voice telephone network accessible to all those with telephones and access privileges (i.e. In the U.S., it was formerly called the Bell System network or the AT&T long distance network).

PULSE DIALING
One or two types of dialing that uses rotary pulses to generate the telephone number.

PULSE TO TONE CONVERTER
A device which recognizes the "clicks" made by a rotary dial phone and converts them to DTMF tones. Not always a reliable technology. Consider the problem faced by the device in distinguishing the click made by dialing a "1" digit and a static click caused by lightning or other interference on the line.

QAM
Quadrature Amplitude Modulation. A modulation technique that uses variations in signal amplitude, allowing data-encoded

symbols to be represented as any of 16 or 32 different states. Some QAM modems allow dial-up data rates of up to 9,600 bps.

QUEUING

The act of "stacking" or holding calls to be handled by a specific person, trunk, or trunk group.

RAM

Random Access Memory. The primary memory in a computer. Memory that can be overwritten with new information. The "random access" part of the name comes from the fact that the next bit of information in RAM can be located no matter where it is, in an equal amount of time, making access to it considerably faster than to information in other storage media, such as a hard disk.

RASTER SCANNING

The method of scanning in which the scanning spot moves along a network of parallel lines, either from side to side or top to bottom.

REFERENCE LINE

The first scanning line in memory. The location of each black pixel of this line is kept in memory for the next scanned line. Depending on the compression technique used, more or fewer scan lines are necessary.

RESOLUTION

A measure of capability to delineate picture detail.

RESOLUTION LEVELS

Provides for high- and low-resolution levels, (the ITU-TSS Group 3 standard is most popular) up to 400 dots per inch. Most computer fax products allow users to select which resolution meets their needs.

RESOURCE MODULE

A Dialogic term referring to devices which perform specific voice processing functions such as voice compression, voice recognition, facsimile transmission and reception, and conversion of computer text to spoken words over the telephone. Resource modules are typically connected to the telephone network through network interface modules.

RING

1. As in Tip and Ring. One of the two wires (the two are Tip and Ring) needed to set up a telephone connection. 2. Also a reference to the ringing of the telephone set. 3. The design of a Local Area Network (LAN) in which the wiring loops from one workstation to another, forming a circle (thus, the term "ring"). In a ring LAN, data is sent from workstation to workstation around the loop in the same direction. Each workstation (which is usually a PC) acts as a repeater by re-sending messages to the next PC in the ring. The more PC's, the slower the LAN. Network control is distributed in a ring network. Since the message passes through each PC, loss of one PC may disable the entire network. However, most ring LANs recover very quickly should one PC die or be turned off. If it dies, you can remove it physically from the network. If it's off, the network senses that and the token ignores that machine. In some token LANs, the LAN will close around a dead workstation and join the two workstations on either side together. If you lose the PC doing the control functions, another PC will jump in and take over. This is how the IBM Token-Passing Ring works.

RJ-11

RJ-11 is a six conductor modular jack that is typically wired for four conductors (i.e. four wires). Occasionally it is wired for only two conductors C especially if you're only wiring up for tip and ring. The RJ-11 jack (also called plug) is the most common telephone jack in the world. The RJ-11 is typically used for connecting telephone instruments, modems and fax machines to a female RJ-22 jack on the wall or in the floor. That jack in turn is connected to twisted wire coming in from "the network" C which might be a PBX or the local telephone company central office. RJ-22 wiring is typically flat. None of its conductors (i.e. wires) are twisted. You cannot use flat cable for high-speed data communications, like local area networks. See also RJ-22 and RJ-45.

RJ-14
A jack that looks and is exactly like the standard RJ-11 that you see on every single line telephone. Whereas the RJ-11 defines one line C with the two center, red and green, conductors being tip and ring, the RJ-14 defines two phone lines. One of the lines is the "normal" RJ-11 line C the red and green center conductors. The second line

ROUTING
The process of selecting the correct path for a message.

RS-232
A set of standards, specifying the various electrical and mechanical characteristics for interfaces between computers, terminals, and modems. It applies to synchronous and asynchronous binary data transmission.

RTN
Retrain Negative.

RTP
Retrain Positive.

SAME-CALL FAX
See ONE-CALL FAX.

SCANNER
A device used to input graphic images in digital form. A fax machine's scanner determines the brightness of a document's pixel for transmission.

SCHEDULED TRANSMISSION
A feature allowing the user to schedule a fax transmission at a specific date or time in the future. Key benefits are convenience and cost savings. Scheduling jobs at a period of low telephone rates can have immediate considerable savings.

SC BUS
The next generation bus now under development by Dialogic and others. See SCSA. Will play a role analogous to the PEB.

SCRIPT

1. A written document specifying the wording of menus and informational messages to be recorded when designing a voice response application. 2. The flow-chart or other description specifying the way that a voice response system interacts with a caller.

SCSA

(pronounced "scuzza") Signal Computing System Architecture. SCSA is a standard for all levels of design of voice processing components from the voice bus chip level to the applications programming interface (API).

SCSA is an open standard. A consortium of leading telecommunications and computing technology players, led by Dialogic and including companies such as IBM and Seimens are, at the time of writing, cooperating in development of the SCSA specification. The specification documents will be available to anyone who wants them for a nominal fee. There will be no technology license fees charged to developers who wish to create SCSA-compatible products.

The most important components of SCSA for the voice processing developer are a bus and a uniform API.

The primary bus is the SCbus. SCbus is a high capacity bus which is designed to be the "next generation PEB". Where the PEB has up to 32 time-slots, the SCbus will have up to 2,048 time-slots: enough band-width for high-fidelity audio, full-motion video and other demanding applications of the future. The standardized API should make SCSA components such as voice processing, voice recognition, speech synthesis, video boards and others accessible independent of the component manufacturer.

SERVER

A computer providing a service such as shared access to a file system, a printer or an E-mail system to LAN users. Usually a combination of hardware and software. A process in a distributed computing system that provides a service in response to requests from clients.

SDLC

Synchronous Data-Link Control.

SPOOLER

A program that controls spooling. Spooling, a term mostly associated with printers, stands for Simultaneous Peripheral Operations On Line. Spooling, temporarily stores programs or program outputs on magnetic tape, or disks for output or processing. On a LAN, a printer is controlled by a spooler. The spooler places each print request on the LAN in the print queue and prints it when it reaches the top of the queue.

STANDARDS

Agreed principles of protocol. Standards are set by committees working under various trade and international organizations. RS standards, such as RS-232-C are set by the "EIA", the Electronics Industries Association. "ANSI" standards for data communications are from the X committee. Standards from ANSI would look like X3.4-1967, which is the standard for the ASCII code. The ITU-T does not put out standards, but rather publishes recommendations, owing to the international personalities and countries involved. "V" series recommendations refer to data transmission over the telephone network, while "X" series recommendations, such as X.25, refer to data transmission over public data networks.

SWITCHED NETWORK

A network providing switched communications service; that is, the network is shared among many users, any of whom may establish communication between desired points when required.

SWITCH

A mechanical, electrical or electronic device which opens or closes circuits, completes or breaks an electrical path, or selects paths or circuits.

T-1

Also spelled T1. A digital transmission link with a capacity of 1.544 Mbps (1,544,000 bits per second). T-1 uses two pairs of normal twisted wires, the same as you'd find in your house. T-1 normally can handle 24 voice conversations, each one digitized at 64 Kbps. T-1 is a standard for digital transmission in North America. It is usually provided by the phone company and used for connecting networks across remote distances. Bridges and routers are used to connect LANs over T-1 networks. In Europe the similar but incompatible service is called E-1 or E1. See A & B BITS for details of signaling.

For a full explanation of T1 see Bill Flanagan's book The Guide to T-1 Networking. (Call 1-800-LIBRARY for your copy.)

T-2

A digital transmission link digital transmission link with a capacity of 6.312 Mbps. T-2 can handle at least 96 voice messages.

T-3

A digital transmission link with a capacity of 44.736 Mbps. T-3 can handle 672 voice messages simultaneously. T-3 runs on fiber optic cable.

T.4

ITU-T recommendation for Group 3 devices, providing definitions of various V-series protocols and signals used during Group 3 operations, including: all supported resolutions, one-dimensional encoding and two-dimensional encoding, and optional error control and error-limiting modes.

T.6

ITU-T recommendation for Group 4 machines. It defines the facsimile coding schemes and their associated coding control functions for black and white images.

T.30

ITU-T recommendation. This handshake protocol describes the overall procedure for establishing and managing communication between two fax devices. It covers five phases of operation: call setup, pre-message procedure (selecting the communication mode), message transmission (including both phasing and synchronization), post-message procedure (EOM and confirmation), and call release.

T.35

ITU-T recommendation proposing a procedure for the allocation of ITU-T members' country or area code for non-standard facilities in telematic services.

T.611

Also known as Appli/COM. A messaging standard proposed by France and Germany, defining a programmable communication interface (PCI) for Group 3 fax, Group 4 fax, teletext, and telex service. It provides communications between the local application and the communications application. This local application to communications application relationship (client/server) exists in several places in the protocol stack.

TCF

Training Check Frame. Last step in a series of signals called a training sequence, designed to let the receiver adjust to line conditions.

TELEFAX

European term for fax.

TELECOPIER

European term for fax.

TEXT-TO-SPEECH

Technology for converting speech in the form of ASCII or other text to a synthesized voice.

TIFF FILE FORMAT

Tagged Image File Format. TIFF provides a way of storing and exchanging digital image data. Aldus Corp., Microsoft Corp., and major scanner vendors developed TIFF to help link scanned images with the popular desktop publishing applications. It is now used for many different types of software applications ranging from medical imagery to fax modem data transfers, CAD programs, and 3D graphic packages. The current TIFF specification supports three main types of image data: Black and white data, halftones or dithered data, and grayscale data. A special variant of TIFF, called TIFF/F, has been defined specifically for storing fax images. Note that most standard PC graphics software, at least at the time of writing, doesn't support TIFF/F even it does support other flavors of TIFF file.

TIP

1. The first wire in a pair of wires. The second wire is called the "ring" wire. 2. A conductor in a telephone cable pair which is usually connected to positive side of a battery at the telephone company's central office. It is the phone industry's equivalent of Ground in a normal electrical circuit.

TONE DIAL

A pushbutton telephone dial that makes a different sound (in fact, a combination of two tones) for each number pushed. The correct name for tone dial is "Dual Tone MultiFrequency" (DTMF). This is because each button generates two tones, one from a "high" group of frequencies C 1209, 1136, 1477 and 1633 Hz C and one from a "low" group of frequencies C 697, 770, 852 and 841 Hz. The frequencies and the keyboard, layout have been internationally standardized, but the tolerances on individual frequencies do vary between countries. This makes it more difficult to take a touch-tone phone overseas than a rotary phone.

You can "dial" a number faster on a tone dial than on a rotary dial, but you make more mistakes on a tone dial and have to redial more often. Some people actually find rotary dials to be, on average, faster for them. The design of all tone dials is stupid. Deliberately so. They were deliberately designed to be the exact opposite (i.e. upside down) of the standard calculator pad, now incorporated into virtually all computer keyboards. The reason for the dumb phone design was to slow the user's dialing down to the speed Bell central offices of early touch tone vintage could take. Today, central offices can accept tone dialing at high speed. But sadly, no one in North America makes a phone with a sensible, calculator pad or computer keyboard dial. On some telephone/computer work-stations you can dial using the calculator pad on the keyboard. This is a breakthrough. It a lot faster to use this pad. The keys are larger, more sensibly laid out and can actually be touch-typed (like on a keyboard.) Nobody, but nobody can "touch-type" a conventional telephone tone pad. A tone dial on a telephone can provide access to various special services and features C from ordering your groceries over the phone to inquiring into the prices of your (hopefully) rising stocks.

TOUCH TONE
A former trademark once owned by AT&T for a tone used to dial numbers. For a full explanation of touchtone, see DTMF.

TRAINING SEQUENCE
Part of the hand-shake used in establishing a fax call where the two devices can adjust to prevailing line conditions.

TRELLIS CODING
A method of forward error correction used in certain high-speed modems where each signal element is assigned a coded binary value representing that element's phase and amplitude. It allows the receiving modem to determine, based on the value of the preceding signal, whether or not a given signal element is received in error.

TRELLIS CODING MODULATION (TCM)
A version of quadrature amplitude modulation that enables relatively high bit rate signals to be used on ordinary voice-grade circuits. Used as a modem modulation technique in which algorithms are used to predict the best fit between the incoming signal and a large set of possible combinations of amplitude and phase changes. TCM provides for transmission speeds of 14.4 kbps and above on single voice-grade telephone lines.

TRUNK
A communication line between two switching systems. The term "switching system" typically includes the equipment in a

telephone company's central office and PBXs. A "tie trunk" connects PBXs. Central office trunks connect a PBX to the switching system at the central office.

TSI
Transmitting Subscriber Information. A frame that may be sent by the caller, with the caller's telephone number (may be used to screen calls).

TSR
Terminate and Stay Resident. A term for loading a software program in a DOS computer in which the program loads into RAM and is always ready for running at the touch of a combination of keys.

TSS
See CCITT.

TTI
Transmit Terminal Identification. The telephone number and words on top of a received fax document, identifying its point of origin. This information does not originate with the telephone company, but with the sender, who programs it into the fax machine.

TWISTED PAIR
Two insulated copper wires twisted around each other to reduce induction (thus interference) from one wire to the other. The twists, or lays, are varied in length to reduce the potential for signal interference between pairs. Several sets of twisted pair wires may be enclosed in a single cable. In cables greater than 25 pairs, the twisted pairs are grouped and bound together in a common cable sheath. Twisted pair cable is the most common type of transmission media. It is the normal cabling from a central office to your home or office, or from your PBX to your office phone. Twisted pair wiring comes in various thicknesses. As a general rule, the thicker the cable is, the better the quality of the conversation and the longer cable can be and still get acceptable conversation quality. However, the thicker it is, the more it costs.

TWO-DIMENSIONAL CODING

A data compression scheme that uses the previous scan line as a reference when scanning a subsequent line. Because an image has a high degree of correlation vertically as well as horizontally, 2-D coding schemes work only with variable increments between one line and the next, permitting higher data compression.

V.17

ITU-T recommendation for simplex modulation technique for use in extended Group 3 facsimile applications only. Provides 7,200, 9,600, 12,000, and 14,400 bps trellis-coded modulation.

V.21

ITU-T recommendation for 300 bps duplex modems for use on the switched telephone network. V.21 modulation is used in a half-duplex mode for Group 3 fax negotiation and control procedures.

V.22

ITU-T recommendation for 1,200 bps duplex modem for use on the switched telephone network and on leased lines.

V.22bis

ITU-T recommendation for 2,400 bps duplex modems for use on the switched telephone network. V.22 also provides for 1,200 bps operation for V.22 compatibility.

V.27ter

ITU-T recommendation for 2.4/4.8-kbps/s modem for use on the switched telephone network. Half-duplex only. It defines the modulation scheme for Group 3 facsimile for image transfer at 2,400 and 4,800 bps.

V.29

ITU-T recommendation for 9,600 bps modem for use on point-to-point leased circuits. This is the modulation technique used in Group 3 fax for image transfer at 7,200 and 9,600 bps. V.29 uses a carrier frequency of 1,700 Hz which is varied in both phase and amplitude. V.29 can be full duplex on four-wire leased circuits, or half duplex on two-wire and dial-up circuits.

V.32

ITU-T recommendation for 9,600 bps two-wire full duplex modem operating on regular dial-up lines or two-wire leased lines. V.32 also provides fallback operation at 4,800 bps.

V.32bis

ITU-T recommendation for full-duplex transmission on two-wire leased and dial-up lines at 4,800, 7,200, 9,600, 12,000, and 14,400 bps. Provides backward compatibility with V.32. It includes a rapid change renegotiation feature for quick and smooth rate changes when line conditions change.

V.33

ITU-T recommendation for 14.4 kbps and 1.2 kbps modem for use on four-wire leased lines.

V.42

ITU-T recommendation, primarily concerned with error correction and compression modems. The protocol is designed to detect errors in transmission and recover with a retransmission.

V.fast

A modem standard under development. Once approved by the ITU-T, it will raise modem speeds to 19.2 kbps.

VOICE BOARD

Also called a voice card or speech card. A Voice Board is a computer add-in card which can perform voice processing functions. A voice board has several important characteristics: It has a computer bus connection. It has a telephone line interface. It typically has a voice bus connection. And it supports one of several operating systems, e.g. MS-DOS, UNIX. At a minimum, a voice board will usually include support for going on and off-hook (answering, initiating and terminating a call); notification of call termination (hang-up detection); sending flash hook; and dialing digits (touchtone and rotary). See VRU.

VOICE INTEGRATION

Allows computer fax solutions to be store-and-forward hubs for both image as well as voice communication. Many of these products work on PC-based systems and offer all the capabilities of a message center.

VOICE MAIL

You call a number. A machine answers. "Sorry. I'm not in. Leave me a message and I'll call you back." It could be a $50 answering machine. Or it could be a $200,000 voice mail system. The primary purpose is the same C to leave someone a message. After that, the differences become profound. A voice mail system lets you handle a voice message as you would a paper message. You can copy it, store it, send it to one or many people, with or without your own comments. When voice mail helps business, it has enormous benefits. When it's abused C such as when people "hide" behind it and never return their messages C it's useless. Some people hate voice mail. Some people love it. It's clearly here to stay.

VOICE MESSAGING

Recording, storing, playing back and distributing phone messages. New York Telephone has an interesting way of looking at voice messaging. NYTel sees it as four distinct areas: 1. Voice Mail, where messages can be retrieved and played back at any time from a user's "voice mailbox"; 2. Call Answering, which routes calls made to a busy/no answer extension into a voice mailbox; 3. Call Processing, which lets callers route themselves among destinations via their touch-tone phones; and 4. Information Mailbox, which stores general recorded information for callers to hear.

VOICE RECOGNITION

The ability of a machine to understand human speech. When applied to telephony environments, the limited bandwidth (range of frequencies transmitted by a telephone connection) and other factors such as background noise and the poor quality of most telephone microphones severely limits the ability of current technology to recognize spoken words. Typical systems are able to recognize standard vocabularies of 16 or so words, such as the digits, yes, no and stop.

VOICE RESPONSE UNIT (VRU

Think of a Voice Response Unit (also called Interactive Voice Response Unit) as a voice computer. Where a computer has a keyboard for entering information, an IVR uses remote touch-tone telephones. Where a computer has a screen for showing the results, an IVR uses a digitized synthesized voice to "read" the screen to the distant caller. An IVR can do whatever a computer can, from looking up train timetables to moving calls around an automatic call distributor (ACD). The only limitation on an IVR is that you can't present as many alternatives on a phone as you can on a screen. The caller's brain simply won't remember more than a few. With IVR, you have to present the menus in smaller chunks. See VOICE BOARD.

VRU

See VOICE RESPONSE UNIT.

WAN

Wide Area Network. A network using common carrier-provided lines that cover an extended geographical area. WANs are data networks typically extending a LAN outside the building, over telephone common carrier lines to link to other LANs in remote buildings or other geographic areas.

WINK

A signal sent between two telecommunications devices as part of a "hand-shaking" protocol. On a digital connection such as a T-1 circuit, a wink is signaled by a brief change in the A and B signaling bits from off to on and back to off (the reverse of a flash-hook). On an analog line, a wink is signaled by a change in polarity (electrical + and -) on the line.

WINK OPERATION

A timed, off-hook signal, normally of 140 ms, which indicates the availability of an incoming register for receiving digital information from the calling office.

WINK START

Short duration off-hook signal. (See Wink Operation.)

WYSIWYG

(Pronounced: Wiz-E-Wig) What You See Is What You Get. You could argue this has nothing to do with fax (it's remote), but I like it anyway.

X.25

Possibly one of the most important int'l standards ever recommended by the ITU-T. From the beginning, it has provided a common reference point by which mainframe computers, word processors, mini-computers, VDUs, microcomputers, and varied equipment from different manufacturers can operate together over a packet switched network. X.25, defines the interface between a public data network and a packet-mode user device. "X" stands for "packet switched network."

X.38

ITU-T recommendation for access of Group 3 facsimile equipment to the Facsimile Packet Assembly/Disassembly (FPAD) facility in public data networks situated in the same country.

X.39

ITU-T recommendation for the exchange of control information and user data between a Facsimile Packet Assembly/Disassembly (FPAD) facility and a packet mode data terminal equipment (DTE) or another pad, for international internetworking.

X.400

ITU-T recommendations for the transmission of electronic text and graphic mail between unlike computers, terminals, and computer networks. Briefly, X.400 describes how mail messages are encoded, and X.25 sets up how they are transmitted.

X.500

The ITU-T standard defining directory services, most commonly for email systems. It provides for common naming and addressing in the networks of large-scale enterprises.

ZERO FILL

A traditional fax device is mechanical. It must reset its printer and advance the page as it prints each scan line it receives. If the receiving machine's printing capability is slower than the transmitting device's data sending capability, the transmitting device adds "fill bits" to pad out the span of send time, giving the slower remote machine the additional time it needs to reset prior to receiving the next scan line.

Index

The Kauffmann Group

Founded in 1990, The Kauffman Group Inc is an enhanced fax consulting firm, focused on the sales, marketing and communications benefits of facsimile technology.

As managing partner, Maury Kauffman is the definitive fax evangelist. He has confidentially advised many of the fax industry's most recognizable companies.

Respected as one of the world's leading authorities on enhanced fax, Kauffman is a contributor to Computer Telephony magazine. He has also written over 40 articles for such publications as *Information Week*, *VAR Business*, *NetWare Solutions*, *Sales and Marketing Strategies* and *Voice Asia*. Kauffman has been quoted in publications ranging from *Business Week* to *Web Week* and from *TeleTalk* (Germany) to the *South China Morning News*. This is Kauffman's second book about Fax.

Kauffman is a perennial speaker at technology and communications conferences worldwide. Besides every Computer Telephony Conference, Kauffman has been a featured speaker at: NetWorld+Interop, Internet World, ISPCon, CommUnity, Voice on the Net, FaxWorld, Fax Directions and the Direct Marketing Association's Annual Conference. Kauffman was the only US-based consultant invited to speak at the first ASIAFAX conference, held in Hong Kong. He addressed the European Association of Newspaper and Magazine Publishers at their annual conference in Zurich. Kauffman was Chairman of the FAXASIA conference, held in Singapore.

In December, 1995, Kauffman was awarded a Star of the Industry by CT magazine.

Clients of The Kauffman Group include: AT&T, AVT/RightFax, Brooktrout Technology, Castelle/Ibex Technologies, CFAX, Dialogic/GammaLink, Epigraphx, FaxSav, IBM, IBM-ISSC, IBM-UK, Parity Software Development Company, Premiere Technologies/Xpedite Systems, Sprint, and US West.

Special Offer

Would you like you company, product, application or story highlighted in my next book? Send me your literature AND add me to your mailing list!

Want FREE INFORMATION and UPDATES?

Want to learn more?

Want to hear about new fax applications?

Be Added to my Fax Broadcast and Mailing Lists!

Send your name, address, phone and Fax to:

Maury Kauffman
Managing Partner
The Kauffman Group
324 Windsor Drive
Cherry Hill, NJ 08002
609-482-8288
609-482-8940 Fax

Maury@KauffmanGroup.com

www.KauffmanGroup.com

SIMPLY, Fax your business card to: 609-482-8940 and write "Mailing List," I'll do the rest.